The Concept of a Real-Time Enterprise in Manufacturing

Daniel Metz

The Concept of a Real-Time Enterprise in Manufacturing

Design and Implementation of a Framework based on EDA and CEP

 Springer Gabler

Daniel Metz
Siegen, Germany

Dissertation Universität Siegen 2013

ISBN 978-3-658-03749-9   ISBN 978-3-658-03750-5 (eBook)
DOI 10.1007/978-3-658-03750-5

The Deutsche Nationalbibliothek lists this publication in the Deutsche Nationalbibliografie; detailed bibliographic data are available in the Internet at http://dnb.d-nb.de.

Library of Congress Control Number: 2013949425

Springer Gabler
© Springer Fachmedien Wiesbaden 2014

Printed on acid-free paper

Springer Gabler is a brand of Springer DE.
Springer DE is part of Springer Science+Business Media.
www.springer-gabler.de

To Mom and Dad.

Foreword from Academia

The subject of the dissertation of Mr. Daniel Metz at the Business & Information Systems Engineering (BISE) Institute at the University of Siegen is the analysis surrounding the concept of real-time enterprise (RTE), and supporting technologies in the last decade, with the main intention to identify shortcomings. Subsequently, Mr. Metz has developed a reference architecture that overcomes temporal and semantic vertical integration gaps, which is crucial to realize the concept of an RTE, across different enterprise levels by exploiting the paradigm of event-driven architecture (EDA) and complex event processing (CEP). The developed reference architecture has been implemented and validated in a foundry, which has typical characteristics of small- and medium-sized enterprises (see the following foreword from industry).

The reference architecture integrates the real-time process data from different resources located on the shop floor with the corresponding (offline) transactional data from the enterprise applications, like an enterprise resource planning (ERP) system. The integrated data is utilized in manifold ways–computation of key performance indicators (KPIs), visualization to enterprise members, creation of logistic information (i.e., tracking and tracing of different enterprise entities), and analysis of the integrated data/event streams with the support of a CEP engine. Especially, CEP is employed to identify any deviations between the planned data and actual values. In addition to these classical reactive/feedback approaches, the reference architecture also supports proactive approaches to enhance the performance of an enterprise.

Further, the reference architecture has been implemented and validated in a state-of-the-art aluminum sand casting enterprise. The implementation utilizes modern methods and techniques of software engineering (e.g., .NET framework, EsperTech CEP engine). Additionally, the implementation aligns with numerous national and international standards and models (e.g., IEC 62264, MESA Model).

The book may contribute to further understanding and development of the concept and realization of RTE and CEP.

Siegen, August 2013 *Prof. Dr.-Ing. Dr. h.c. Manfred Grauer*
(University Siegen)

Foreword from Industry

Compared to many other manufacturing processes, the casting process offers a high level of free design options. Even for complex product geometries a near-net-shape-production can be achieved in highly automated production lines. From the numerous degrees of freedom in the casting process derives a large variety of very different impact parameters, which form a very complex relationship.

In 2008, the company Ohm & Häner opened a state-of-the-art production facility for aluminum sand casting based on clay-bonded sand molds. All relevant production systems within the plant are directly accessible by IT systems. Further, it was part of the planning concept to integrate an online data management system. Therefore, the AutoEDA project has been initiated as a bilateral research project between the company Ohm & Häner and the BISE Institute at the University of Siegen.

In the years 2009 to 2013, Mr. Daniel Metz has intensively engaged himself with the sand casting process and considered it by the methods of IT. It was an essential element of his research to comprehensively collect the process parameters from the manufacturing systems/resources in near real-time and to evaluate the data directly in terms of consistency with the requirements of the company's planning systems.

Starting from the outline of a complex process model, Mr. Metz created an appropriate IT framework that enables the company to consider online the various conditions of the production, the critical process parameters and other relevant data in a general context. Thus, critical situations can immediately be identified and eliminated in the running production. Moreover, the system offers the possibility to directly generate key figures of the production performance and report the information to various divisions of the company for further analysis and process optimization. The existing software solution resulting from the scientific work of Mr. Metz is now an essential and convenient tool for the daily production routine of the foundry with regard to both the optimization of technical processes and cost reduction.

Olpe, August 2013

Dr.-Ing. Georg Dieckhues
Dr.-Ing. Ludger Ohm
(Ohm & Häner Metallwerk GmbH & Co. KG)

Acknowledgements

I consider it as a gift and grace from God to got the chance to do my research and complete my Ph.D. at the Business & Information Systems Engineering (BISE) Institute at the University of Siegen, Germany. This was absolutely out of range of what I could do on my own. "My grace is sufficient for you, for my power is made perfect in weakness", 2. Corinthians 12:9. Thus, first of all, I want to praise and thank God for all the blessings I have received with his love. In the name of his son *Jesus*

Soli deo gloria.

Further, I want to thank my *mom and dad* for their entire love and support they have given to me. I am very thankful to my *family*–inclusive my sister and brother– that has given me all the time so that I could fully concentrate on my studies and work. Moreover, I appreciate all encouraging words that have built me up when I was down. In that, I am also grateful to my wife *Katharina* for her love. Thank you, my darling, for your support on evenings and weekends where I had to work on finalizing this book. I love you.

Several people have partly joined my journey at the BISE institute and have inspired and supported my work in different ways. My colleague and close friend *Ulf Müller* is the one that has accompanied me for more than ten years till today. I went with him through all the ups and downs you usually have during (Ph.D.) studies. Thank you very much my friend.

In the beginning of my time at the BISE institute, I contributed to the BEinGRID project and I have learned a lot from my former colleague *Dr. Julian Reichwald* who was an inspiring example for me at the BISE institute.

Then, I was part of the AutoEDA project that had been started together with Ohm & Häner Metallwerk GmbH & Co. KG, Olpe, Germany. I appreciate the work of my former colleague *Walter Schäfer* who contributed to the project in the beginning. Later, *Sachin Karadgi* joined the project and I spent many hours–sometimes it was really too late–together with him in the plant and doing the implementation. Most of my research papers have been written together with *Sachin* who became a close friend of mine. I have learned many things from him, not only technically, but also for life. I can honestly proclaim that the success of the AutoEDA project is closely related to the excellent teamwork I had with *Sachin*.

In addition, the realization of the AutoEDA project was only possible because of our industrial partner Ohm & Häner Metallwerk GmbH & Co. KG, Olpe, Germany. This company had the initial vision and the courage to start the three years research and development project AutoEDA. Namely, I am thankful to *Dr.-Ing. Ludger Ohm*, *Dr.-Ing. Georg Dieckhues* and *Jürgen Alfes* for the remarkable and outstanding cooperation. Especially, *Dr.-Ing. Georg Dieckhues* has done an excellent job in managing the project and providing me with all support that was necessary. Also, he organized and hosted many presentations of the AutoEDA system to external (industrial) partners. Moreover, I wish to thank all members of the Ohm & Häner Metallwerk GmbH & Co. KG for their valuable feedback and contribution in the project. This entails the IT department with *Jürgen Alfes, Sebastian Rath, Ulrich Reuber*, and *Jürgen Pulte* as well as the workers on the shop floor.

Also, I am greatful to *Matthias Dittrich* from the Heinrich Wagner Sinto (HWS) Maschinenfabrik GmbH, Bad Laasphe, Germany. He has fostered my insights into the shop floor (i.e., machines, automation systems and programmable logic controllers) and delivered relevant information and ideas to connect shop floor resources. In addition, we had a great cooperation–and a lot of fun, too–in the MOLD-CONTROL project.

Some students have done their master thesis during the AutoEDA project, and I have had the honor to guide and support them in doing so. The project gained from the work of these students. Thus, I thank *Markus Weber, Philipp Koppmann, Benjamin Petri, Andreas Mailinger*, and *Lars Utermöhlen* for their outstanding project work. Further, I want to thank *Marco Schneider* for his great job working on the EsperTech complex event processing (CEP) engine.

Last but not least, I am very greatful to my advisor *Prof. Dr.-Ing. Dr. h.c. Manfred Grauer*. I appreciate the fruitful discussions concerning the concept of real-time enterprise (RTE), closed-loop controls, and the like. In addition, I want to thank all (former) staff members at the BISE institute, especially, *Annette Wiebusch* and *Ralf Dreier* for supporting me. Finally, I thank my second reviewer *Prof. Dr. Stein* for his suggestions pertaining to the philosophy of sciences.

Langenargen, August 2013   *Daniel Metz*

Contents

List of Figures

Acronyms

ADL   Architecture Description Language
AI   Artificial Intelligence
ARIS   Architecture of Integrated Information Systems
BAM   Business Activity Monitoring
BEMN   Business Event Modeling Notation
BFU   Basic Fractal Unit
BMBF   Bundesministerium für Bildung und Forschung
BMS   Bionic Manufacturing System
BMWi   Bundesministerium für Wirtschaft und Technologie
BOM   Bill Of Materials
BPM   Business Process Management
BPML   Business Process Modeling Language
BPMN   Business Process Model Notation
BPR   Business Process Reengineering
CEP   Complex Event Processing
CER   Constrained Execution Region
CIM   Computer Integrated Manufacturing
CIMOSA   CIM Open System Architecture
CIP   Continual Improvement Process
CLR   Common Language Runtime
CONWIP   Constant Work In Progress
CORBA   Common Object Request Broker Architecture
CPS   Cyber-Physical System
CPU   Central Processing Unit
CRM   Customer Relationship Management
DA   Data Access
DAG   Directed Acyclic Graph
DBMS   Database Management System
DCOM   Distributed Component Object Model
DEBS   Distributed Event-Based System
DES   Discrete Event Simulation

DFA   Deterministic Finite Automaton
DFD   Data Flow Diagram
DPWS   Devices Profile for Web Services
DSMS   Data Stream Management System
EAI   Enterprise Application Integration
ECA   Event Condition Action
EDA   Event-Driven Architecture
EDBPM   Event-Driven Business Process Management
EDG   Event Detection Graph
EI   Enterprise Integration
EP   Event Processing
EPA   Event Processing Agent
EPC   Event-Driven Process Chains
EPL   Event Processing Language
EPN   Event Processing Network
EPTS   Event Processing Technical Society
EQL   Event Query Language
ERP   Enterprise Resource Planning
ESP   Event Stream Processing
EU   European Union
FSM   Finite State Machine
GERAM   Generalized Enterprise Reference Architecture and Methodology
GIM   GRAI Integrated Methodology
GPSS   General Purpose Simulation System
GUI   Graphical User Interface
HANA   High-Performance Analytical Appliance
HDL   Hardware Description Language
HMS   Holonic Manufacturing System
ICT   Information & Communication Technology
IDE   Integrated Development Environment
IEC   International Electrotechnical Commission
IfM   Institut für Mittelstandsforschung
IMS   Intelligent Manufacturing System
IoT   Internet of Things
IP   Internet Protocol
IPA   Fraunhofer-Institut für Produktionstechnik und Automatisierung
ISA   The International Society of Automation
ISO   International Organization for Standardization
ISR   Information Systems Research
IT   Information Technology
JEE   Java Platform, Enterprise Edition
JIS   Just In Sequence
KDD   Knowledge Discovery in Database
KMDL   Knowledge Management Description Language
KPI   Key Performance Indicator

LDAP	Lightweight Directory Access Protocol
MAS	Multi-Agent System
MDA	Model-Driven Architecture
MES	Manufacturing Execution System
MESA	Manufacturing Enterprise Solutions Association
MIL	Module Interconnection Language
MIT	Massachusetts Institute of Technology
MOM	Manufacturing Operations Management
MOM	Message-Oriented Middleware
MPS	Master Production Schedule
MRP	Material Requirements Planning
MRP II	Manufacturing Resource Planning
MTO	Make-To-Order
NC	Numerical Control
NFA	Non-Deterministic Finite Automaton
NSF	National Science Foundation
OCE	Online Control Engine
OEE	Overall Equipment Effectiveness
OLE	Object Linking and Embedding
OMG	Object Management Group
OOP	Object-Oriented Programming
OPC	OLE for Process Control
OR	Operations Research
OSI	Open Systems Interconnection
PAIS	Process Aware Information System
PDM	Product Data Management
PERA	Purdue Enterprise Reference Architecture
PLC	Programmable Logic Controllers
PLM	Product Lifecycle Management
POA	Point-Of-Action
POC	Point-Of-Creation
POJO	Plain Old Java Object
PPC	Production Planning and Control
PRM	Purdue Reference Model
RBS	Rule-Based System
RFID	Radio Frequency Identification
RPC	Remote Procedure Call
RTB	Real-Time Businesses
RTE	Real-Time Enterprise
SCM	Supply Chain Management
SEP	Simple Event Processing
SIRI	Service Interface for Real Time Information
SME	Small and Medium-Sized Enterprise
SOA	Service-Oriented Architecture
SOAP	Simple Object Access Protocol

SPC	Statistical Process Control
SQC	Statistical Quality Control
SQL	Structured Query Language
TCP	Transmission Control Protocol
TUM	Technical University of Munich
UA	Unified Architecture
UDDI	Universal Description, Discovery and Integration
UDIT	Unique Device Identification Tag
UML	Unified Modeling Language
UPnP	Universal Plug and Play
USA	United States of America
VDI	Verein Deutscher Ingenieure
VDMA	Verband Deutscher Maschinen- und Anlagenbau
WCF	Windows Communication Foundation
WEF	WSDM Event Format
WS-BPEL	Web Service Business Process Execution Language
WSDL	Web Service Description Language
WSDM	Web Services Distributed Management
XML	eXtensible Modeling Language

Abstract

Not just since the financial crisis in 2008, have manufacturing enterprises been confronted with increasing challenges. Besides factors like quality and cost, time and innovative capabilities (e.g., product design) have become critical success factors in global markets. Enterprises have to react instantly to changing market conditions and disturbances that occur during execution of value creation processes. Depending upon the processes' context, the goal is to significantly reduce lead times, reaction times, and time-to-market, among others. The vision of a real-time enterprise (RTE), which is able to sense and analyze events from internal and external sources, and perform adequate (re-) actions, has been envisaged by manufacturing enterprises. Here, the realization of an RTE has to incorporate management, engineering and computer science perspectives.

The main objective is the realization of closed-loop controls using feedback among the processes' targets (e.g., quality, cost, quantity) defined by responsible process owners and the actual manufacturing process performance at the shop floor level during the execution of value creation processes. Unfortunately, in many instances, the vertical integration of enterprise levels is inadequately implemented. Hence, manufacturing execution systems (MES) try to mitigate the vertical integration gap by introducing feedback among enterprise levels. Nevertheless, the exchange of data is manual or semi-automatic because of the inflexible and proprietary interfaces. Also, process data from automation devices is often not reasonably incorporated. Consequently, the realization of multiple closed-loop controls is still pending.

The concept of an RTE requires information technologies to sense, (pre-) process, and analyze events, and further, deduce appropriate (re-) actions based on detected critical situations. The (software) components of an event-driven architecture (EDA) interact through the exchange of events. Therefore, concepts of an EDA are promising building blocks to implement a vertically integrated (manufacturing) enterprise and establish an RTE. In recent years, the processing of (complex) events, which is an indispensable part of an EDA, has been discussed in the context of complex event processing (CEP).

In this research work a framework based on EDA and CEP is presented towards the realization of RTE in manufacturing. The framework closes the aforementioned vertical integration gap, and further, establishes feedback in (near) real-time among enterprise levels. As such, the framework provides a holistic and closed-loop control of (manufacturing) processes. The framework encompasses results and insights from management, engineering, and computer science. This has been implemented for Ohm & Häner Metallwerk GmbH & Co. KG, Olpe, i.e., a small and medium sized foundry. Here, the framework is employed to monitor and control sand casting processes. The developed control approach has led to a significant increase in (manufacturing) processes' efficiency (i.e., performance, quality, and availability).

Zusammenfassung

Nicht erst mit dem Beginn der Finanzkrise ab 2008 stehen produzierende Unternehmen vor steigenden Herausforderungen. Dabei werden neben den Einflussfaktoren Qualität und Kosten zunehmend auch Zeit und Innovationsfähigkeit zu entscheidenden Erfolgsfaktoren im globalen Wettbewerb. Die Vision eines Echtzeitunternehmens (engl. Real-Time Enterprise (RTE)), das auf interne als auch externe Ereignisse schnell und adäquat reagieren kann, wird daher auch von produzierenden Unternehmen anvisiert. Bei der Etablierung der Vision eines RTE für produzierende Unternehmen sollten neben der betriebswirtschaftlichen Perspektive aber auch ingenieurtechnische und informationstechnische Perspektiven einbezogen werden.

Aus ingenieurtechnischer Perspektive besteht die Zielsetzung im Aufbau zeitnaher Rückkopplungen zwischen den vom Management vorgegebenen Zielgrößen und der tatsächlich erzielten Unternehmensleistung während der Ausführung der Produktionsprozesse. Gegenwärtig ist die vertikale Integration der Unternehmensleitebene mit der Fertigungsebene jedoch nur unzureichend realisiert. Manufacturing Execution Systeme (MES), die u.a. vom Verein Deutscher Ingenieure (VDI) in der Richtlinie VDI 5600 beschrieben werden, versuchen die vertikale Integrationslücke zu schließen und Rückkopplungen zwischen den Unternehmensebenen aufzubauen. Der Austausch von Daten zwischen den Unternehmensebenen erfolgt trotz MES jedoch häufig manuell oder halb-automatisch über inflexible und proprietäre Schnittstellen. Zudem werden Prozessdaten aus den Automatisierungssystemen oftmals nicht in ausreichendem Maße mit einbezogen.

Das Konzept eines RTE setzt aus informationstechnischer Perspektive die Fähigkeit voraus, Ereignisse (– die u.a. auf der Fertigungsebene generiert werden –) zu registrieren, zu verarbeiten und adäquate (Re-) Aktionen auszulösen. In einer ereignisgetriebenen Architektur (EDA) interagieren Komponenten durch den Austausch von Ereignissen (engl. events) miteinander. Das Konzept einer EDA kann daher zur Realisierung der vertikalen Integration und der Etablierung von zeitnahen Rückkopplungen zwischen den Unternehmensebenen herangezogen werden. Die Verarbeitung (komplexer) Ereignisse, die Bestandteil einer EDA ist, wird in jüngerer Vergangenheit unter dem Begriff Complex Event Processing (CEP) diskutiert.

In der Dissertation wird ein auf einer EDA und CEP basierendes Framework vor-gestellt, das die vertikale Integrationslücke schließt und (zeitnahe) Rückkopplungen zwischen den Unternehmensebenen zum Zweck der echtzeitnahen Steuerung der Produktionsprozesse etabliert. Zielsetzung ist die Steigerung der Effizienz der Pro-duktionsprozesse – also der Verbesserung von Durchlaufzeiten, Qualität und Ausla-stung. Das Framework berücksichtigt die oben skizzierten betriebswirtschaftlichen, ingenieurtechnischen und informationstechnischen Perspektiven und wird in Ko-operation mit einer mittelständischen Gießerei (Ohm & Häner Metallwerk GmbH & Co. KG, Olpe) entwickelt und im Kontext der Fertigung hochwertiger Aluminium-Sandgussteile erprobt und evaluiert.

Chapter 1
Introduction

In this chapter, the problem area of the presented research work is motivated, the deduced research goals are defined, the basic positions concerning philosophy of sciences are provided, and the structure of the research work is described.

1.1 Motivation

According to the initial definition of Gartner, a real-time enterprise (RTE) is "an enterprise that competes by using up-to-date information to progressively remove delays to the management and execution of its critical business processes" [60, 1]. The appliance of this vision of an RTE to manufacturing enterprises implies the necessity to monitor and control their value-creation processes in (near) real-time [138]. Thus, a manufacturing enterprise has to (re-) act to both internal events (e.g., changing the priorities within a production schedule) and external events (e.g., demand behavior) to sustain its competitive advantages. Especially, events generated during execution of value creation processes have to be sensed, analyzed and used to manipulate processes expediently. The main aim is a substantial enhancement of the manufacturing processes' efficiency, i.e., improved throughput, quality, and availability [95]. A manufacturing enterprise, which is built in accordance with the aforementioned vision, is defined in the context of this research work as an RTE.

The integration of business functions into a single system using information technology (IT) can be seen as a prerequisite for the realization of the aforesaid (closed-loop) control of value creation processes [156] and the vision of an RTE [60]. Especially for manufacturing enterprises, the challenge is to minimize inconsistencies among the financial, quantitative and qualitative planning and the actual situation during execution of manufacturing processes. This alignment between the desired processes' performance and their factual behavior is mandatory for an integrated enterprise.

At the topmost level of a manufacturing enterprise, i.e., the enterprise control level, enterprise resource planning (ERP) aims at establishing an integrated (IT) system comprising various business functions [94]. To support intercompany processes with an enterprise's suppliers and customers, ERP systems have been extended by supply chain management (SCM) and customer relationship management (CRM) systems [12]. This *horizontal integration* of business processes at the enterprise control level has been fostered by the advent and implementation of the service-oriented paradigm.

In a service-oriented architecture (SOA), business functions are implemented as interoperable services following standards, like web service description language (WSDL) [43]. Hence, services can be employed as building blocks of an enterprise's business processes. Before the concrete implementation of an SOA, business processes have to be consistently (re-) engineered and modeled using (de facto) standards, like the business process model notation (BPMN) (Object Management Group 2011) or event-driven process chains (EPC) [142].

The process models can be manually or even automatically transformed (cf. the work of Reichwald concerning model-driven architectures (MDA) [213]) to service orchestrations[1] (e.g., expressed using the web service business process execution language (WS-BPEL) [13]). Summarized, the horizontal integration of an enterprise at the enterprise control level has traditionally gained a lot of attention in computer science and industry. Not only because of the aforementioned circumstances, the horizontal integration of enterprises has been maturely established [94]. Henceforth, SOA has become a de facto standard for enterprise application integration (EAI), i.e., a means for seamless integration of business functions along the value creation processes.

However, the *actual* value creation processes (i.e., directly added value) are located at the lowest level of a manufacturing enterprise, i.e., at the manufacturing execution or shop floor level. The input factors of these processes are, for instance, materials, manufacturing resources (e.g., machines), and manpower [107]. These input factors have to be combined and transformed to produce (semi-finished) products. In addition to these classical input factors, dispositive factors, like management commitment and context information, have become an indispensable part of today's value creation processes. Specifications regarding a product's bill of materials (BOM) and scheduling of activities have been defined in planning systems (e.g., ERP system). Additionally, the manufacturing processes' performance targets, i.e., production, maintenance, quality, and inventory [247], are pre-defined on higher enterprise levels. Consequently, manufacturing processes should be performed in adherence to the previously mentioned specifications.

In addition to an enterprise's horizontal integration of applying EAI solutions [156, 228], the concept of an RTE in manufacturing consequently requires the *vertical integration* of several enterprise levels [99]. In short, the actual manufacturing process execution has to be monitored and controlled in accordance with the pre-defined processes' specifications and targets.

---

[1] These orchestrated services describe a workflow, i.e., simplified, a part of a business process that can be executed by a process aware information system (PAIS) [2,266]. Often the PAIS is realized as a workflow engine, which is in charge of workflow enactment and control.

Unfortunately, available EAI solutions lack in incorporating manufacturing equipment [94], thus inadequately support (highly) automated manufacturing processes. In addition, SOA is not used extensively to integrate shop floor resources. Although European Union (EU) funded projects have provided prototypical solutions to exploit the SOA paradigm to seamlessly integrate machines and automation systems [237], these implementations are immature and require further development.

Contrary to the predominant request-reply pattern of (traditional) SOA, the vision of an RTE, wherein (multiple) closed-loop controls have to be established, strives to process events using publish-subscribe mechanisms [230]. This type of event processing can be based on concepts of an event-driven architecture (EDA), i.e., an architecture wherein (software) components detect, exchange, and process events [32]. In recent years, event processing has been discussed in the context of complex event processing (CEP). So far, CEP is primarily focused on business and financial processes (e.g., algorithmic trading, business activity monitoring (BAM), telecommunication) [260]. Comprehensive frameworks encompassing process models and implementation guidelines for the appliance of CEP in manufacturing are rare [101].

Apart from the discussion of the above information and event processing paradigms, manufacturing execution systems (MES) have been introduced to bridge the vertical integration gap. In recent years, one has witnessed several industrial organizations (e.g., German Engineering Federation (VDMA)) that have intensified their work on MES [89]. The not-for-profit Manufacturing Enterprise Solutions Association (MESA) deals with standardization and modeling issues of MES [169]. In its current MESA model, it highlights several strategic initiatives supported by vertical integration of an enterprise. Besides strategies, like lean manufacturing and product lifecycle management (PLM), the concept of RTE is mentioned in this model. MES is seen as a prerequisite for an RTE in manufacturing [168].

Although significant progress regarding MES has been achieved, major problems remain open with respect to the interface between the enterprise control level and the manufacturing level [201]. In practice, the exchange of data is done manually or at most semi-automatically because of the inflexible and proprietary interfaces [140]. Because of this inadequate vertical integration of an enterprise, a closed-loop control in (near) real-time is hindered. In addition, comprehensive knowledge management incorporating planned as well as actual process data will be derogated. Consequently, advanced analysis and compilation of (unknown) correlations among process data (e.g., temporal and causal relationships) cannot be accomplished. Finally, an autonomous control of the manufacturing processes by detection of critical process states in real-time, and subsequently, automatic dispatching of (re-) actions is hampered.

To overcome the above shortcomings, several activities have been performed with respect to *standardization*. MESA, for instance, has outlined a master data architecture for manufacturing operations management (MOM) [31]. Also, the introduction of the OLE for Process Control (OPC) UA standard will foster a seamless vertical integration among MES, and automated devices and resources on shop floor [62]. Besides the progress of enhancing communication protocols, like OPC, a draft version of VDI 5600-3 focuses on the *semantics* of machine interfaces and elaborates process data that must be or should be provided by these interfaces [256].

However, in addition to these standardization activities, a framework is required, which concomitantly respects the management, engineering, and information technology perspectives of the aforementioned problem area [95]. The design and implementation of such a framework is typical for the design-oriented information systems research (ISR) as it capitalizes on different technical and non-technical disciplines and aspects. Existing frameworks, which are coping with the vertical integration of a manufacturing enterprise, the realization of holistic monitoring and the control of manufacturing processes, are predominantly based on agent technology (cf. [158]).

On the contrary, frameworks exploiting the EDA paradigm and CEP in manufacturing are rarely addressed in the scientific world (e.g., [101, 265]). So far, the emphasis is on business processes, i.e., on the financial and administrative processes within and across an enterprise. To the author's knowledge, there are *few* MES solution providers that claim to use CEP technology in MES (cf. [91]). However, a concise framework incorporating the management, engineering, and computer science perspectives is missing.

To summarize the above discussion, relevant problem areas surrounding the realization of an RTE in manufacturing are outlined in Fig. 1.1. The definition and formulation of the RTE vision, and the inducement and management of measures for its realization, are primarily in the realm of *enterprise management* (i.e., management community). At a conceptual level, the main challenges regarding the realization of an RTE in manufacturing are the implementation of (i) *enterprise integration*, both horizontally and vertically; and (ii) *real-time monitoring and control* of manufacturing processes. The vertical integration is addressed by *MES*, which is researched, developed, and standardized mainly in the *engineering domain*. Sense-and-respond and multiple control loops can be achieved by various information technologies. Recently, *EDA and CEP* have been primarily discussed in the *computer science domain* as enablers of an RTE that is based on sense-and-respond. As described in pervious paragraphs, one can identify three main problem areas in the context of an RTE in manufacturing. First, standardization activities pertaining to the vertical integration of an enterprise–especially with regard to interfaces between MES and manufacturing resources–have been intensified. Second, real-time monitoring and control of manufacturing processes based on EDA and CEP can be conceptualized, implemented, and evaluated. Finally, EDA and CEP can be combined with MES to realize a (i) seamless enterprise integration; (ii) enhanced agility of MES; and (iii) (multiple) control loops in (near) real-time. The last two problem areas (i.e., problem areas II. and III.) are especially addressed in this research work.

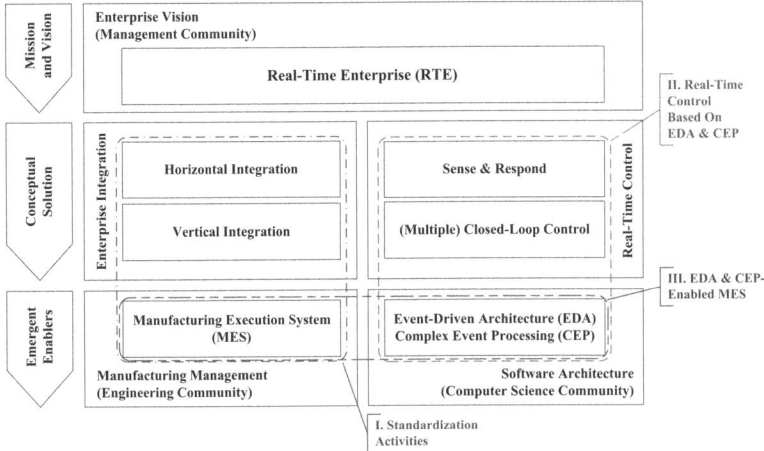

Fig. 1.1 Simplified outline of problem areas surrounding the establishment of an RTE in manufacturing, and adumbration of an inter-disciplinary approach based on EDA and CEP.

1.2 Research Goals

Hence, the aim of this research is the design, implementation and *industrial* evaluation (i.e., evaluation against the considered reality) of a framework based on EDA and CEP, which significantly contributes to the realization of an RTE in manufacturing. This framework encompasses two main modules: (i) a process model that facilitates principal procedures to prepare, implement and introduce the envisioned event-driven architecture in a manufacturing enterprise; and (ii) the event-driven architecture delineating its subcomponents (e.g., CEP), their interconnections, and interplay. The framework is/can be the basis for various functionalities mentioned in the context of manufacturing management (e.g., key performance indicators, genealogy). However, the focus of this research work is on holistic real-time monitoring and control of manufacturing processes employing CEP technology. The problem area regarding standardization activities, like VDI 5600 part 3 (cf. [256]), is taken into account. However, as these standardization activities are largely addressed in the engineering domain, this research work is not intended to substantially contribute to these standardization activities.

1.3 Positioning concerning the Philosophy of Sciences

The presented research work is in line with the tradition of the German *design-oriented ISR* (cf. [199]), yet differs from the *behavioristic approach* favored in the

Anglo-American research community (cf. [214]). The goal of the design-oriented ISR is summarized as to "produce knowledge that helps with developing and deploying information systems successfully" [82, 11], with the intention to design artifacts for *solving* existing real (organizational) *problems* [111, 114, 170, 214]. In addition, Frank argues that ISR even exceeds computer science by designing *new possible worlds* (i.e., incorporation of new ways of organizing work, cooperation and coordination) [82]. Here, the presented research work provides a framework for the realization of the *vision* of an RTE in manufacturing. ISR is not limited to analyzing existing information systems. Rather, especially from a methodological perspective, ISR has some commonalities with engineering disciplines (e.g., mechanical engineering, computer science[2]). The construction of artifacts like information models, frameworks, software prototypes, and so forth, is a main characteristic of the German design-oriented ISR.

The philosophy of sciences has proposed several approaches/guidelines for scientific research encompassing logical positivism, critical rationalism, critical theory, and the like [82]. Frank has analyzed existing approaches in the philosophy of sciences and concludes that none of them "is suited to satisfy the specific requirements of ISR" [82, 21]. The above approaches have evolved in scientific fields, primarily in social and natural sciences, which do not necessarily adhere to the peculiarities of the design-oriented ISR [82]. Natural and social sciences aim at *explaining* the reality, while *design sciences* like engineering and design-oriented ISR try to *shape* the reality by following design goals [114, 170]. Similarly, Simon distinguishes natural sciences and *sciences of the artificial* [234]. Overall, it seems to be impossible to clearly assign the design-oriented ISR, thus also the presented research work, to a classical scientific school. Nevertheless, some overlaps between the presented research work and the general philosophy of sciences are investigated in the following paragraphs.

Becker et al. have presented a framework to disclose the scientific foundation of ISR incorporating ontological, epistemological, and linguistic positions [19]. Similarly, Robra-Bissantz has provided an epistemological framework in the context of ISR [215]. The ontological and epistemological positioning of the presented research work is discussed on the basis of these frameworks.

The presented research work shares the *ontological* position of *realism*, i.e., the world has an objective existence that is independent of the perception of a subject. Further, the implemented framework, the application area/universe of discourse, and their interaction can be perceived by a subject (i.e., human), either directly or indirectly (e.g., assisted by measuring equipment). Consequently, entities and processes of the considered universe of discourse (e.g., manufacturing processes, machines) can be perceived and modeled as process models, software architecture, and the like. The created models are the basis for the implementation of further IT artifacts. In that sense, the *epistemological* position of the presented research work has commonalities with the *critical realism*. It assumes the (limited) capability of a human to perceive the selected slice of reality.

---

[2] However, ISR differs from computer science that "has no direct bearing on organizational or everyday problems" [239, 119].

The presented research work best fits into the *critical rationalism* of Popper based on the above ontological and epistemological positions. According to critical rationalism, hypotheses that can be falsified have to be formulated. The testing of a hypothesis necessitates confronting it with reality [82]. The *hypothesis* that (i) the RTE in manufacturing can be (partly) realized with the presented framework based on EDA and CEP and (ii) the framework contributes to (measurable) improvements (e.g., manufacturing process efficiency expressed as cycle/takt time) can be tested/*evaluated against reality*. However, the evaluation of such an IT artifact is restricted to aspects "regarding the effect of its use" [82, 20]. Furthermore, Frank mentions that the *falsification concept* of critical rationalism is hardly compatible with IT-systems as tests against all cases would be required [82].

Because of the aforementioned difficulties of clearly assigning design-oriented ISR (i.e., also the presented research work) to a classical scientific school, Hevner et al. have described seven guidelines for *design science* in ISR [114]. These guidelines are listed below, and reveal how the presented research work conforms to these guidelines:

- "Design-science research must produce a viable artifact in the form of a construct, a model, a method, or an instantiation" [114, 83]. A framework encompassing a process model and software architecture for implementing an RTE in manufacturing is developed. The software architecture is instantiated/implemented in a foundry (cf. Chap. 5).
- "The objective of design-science research is to develop technology-based solutions to important and relevant business problems" [114, 83]. The concept of an RTE in manufacturing is addressed in the presented research work. This incorporates improvements concerning the horizontal and vertical integration of a manufacturing enterprise as well as monitoring and control of manufacturing processes in (near) real-time.
- "The utility, quality, and efficacy of a design artifact must be rigorously demonstrated via well-executed evaluation methods" [114, 83]. Whether the IT artifact *effectively* satisfies the requirements and constraints of the described problem has been evaluated (cf. Chap. 3). The implemented IT artifact is studied in depth in the intended manufacturing environment (cf. Chaps. 5 and 6), i.e., *observational evaluation* according to Hevner et al. (cf. [114]).
- "Effective design-science research must provide clear and verifiable contributions in the areas of the design artifact, design foundations, and/or design methodologies" [114, 83]. The developed framework, i.e., the process model and the software architecture (cf. Chap. 4), and their implementation (cf. Chap. 5) are contributions in the areas of design artifact and design foundation. The implemented framework contributes to realizing an RTE in manufacturing.
- "Design-science research relies upon the application of rigorous methods in both the construction and evaluation of the design artifact" [114, 83]. The construction of the IT artifact is based on a thorough requirements analysis. Furthermore, established design methods are employed both for the intended manufacturing processes and the software/system architecture. The traits/constraints of the application area/universe of discourse (cf. Chap. 3) have been carefully described.

These traits/constraints are considered to achieve an *abstract* design of the IT artifact. Moreover, design decisions are deduced/justified referring to the knowledge base of the ISR community (e.g., literature).

- Hevner et al. describe design science as an iterative process composed of generating design alternatives and testing of design alternatives against requirements/constraints [114]. The development and implementation of the presented framework have been performed in a similar fashion.

- Finally, Hevner et al. advocate the position that design-science research has to be presented to technology-oriented as well as management-oriented audiences [114]. The framework has been presented to scientific communities in the field of engineering, computer sciences, and management. Moreover, the (implemented) framework has been discussed with practitioners from the intended application area (e.g., foundry industry).

The primary method for acquiring knowledge that is used for/during modeling/design of artifacts in the design-oriented ISR is *deduction* (cf. [199]). Nevertheless, especially during the above mentioned tests of (implemented) design alternatives, *induction* is employed (e.g., analysis of the implemented artifact with, e.g., respect to performance criteria).

Finally, different concepts of truth have been proposed in the philosophy of sciences: (i) correspondence theory; (ii) coherence theory; and (iii) consensus theory. According to the above ontological and epistemological positions, the correspondence theory (i.e., true is what adheres to reality) is applied. In addition, the coherence theory (i.e., true is what fits into a set of true scientific sentences) is used (e.g., literature review).

Because of its roots in both business administration/management and computer science, methodological pluralism is advocated for ISR (cf. [193]). However, the above mentioned design of artifacts like constructs, models, methods, and instantiations (cf. [170]) is a peculiarity of the German ISR. The performed *research process* (cf. [214]) consists of (i) analysis of the application area/universe of discourse and the identification of a research gap (Chaps. 2 and 3); (ii) design of an artifact incorporating requirements/constraints of the application area (Chap. 4); (iii) instantiation/implementation of the artifact, i.e., prototyping [19] (Chap. 5); and (iv) evaluation of the artifact against reality (Chaps. 5 and 6). The design of the artifact is considered in ISR as an *evaluation against the research gap*, while the application of the artifact in the application area is seen as *evaluation against reality* [214]. The latter implies the relevance/adequacy of the research gap [214].

1.4 Structure of the Work

The remainder of the research work is structured as follows. In Chap. 2, an introduction to the research work's problem area is given. The vision and basics of the term RTE are outlined, especially, the application of the RTE vision in manufacturing enterprises. Further, definitions, fundamentals and principles of EDA and CEP are elaborated.

The above mentioned problems can be viewed from different perspectives that are elaborated in Chap. 3. First, the management perspective has its emphasizes on organizational and strategic issues regarding the realization of the RTE in manufacturing. Second, the engineering perspective introduces the concept of feedback control within and across enterprise levels. As a prerequisite for the previously mentioned real-time control, the horizontal and vertical enterprise integration is considered. Subsequently, MES and related standards are described. Third, challenges and requirements for information technology to implement an RTE in manufacturing are investigated. Finally, the requirements gathered from the above perspectives are summarized.

An event-driven framework for the realization of an RTE in manufacturing is described in Chap. 4. Before the outline of this framework, related work concerning intelligent and holistic (real-time) monitoring and control of manufacturing processes is presented. The subsequent presentation of the framework is split into a process model, and an architecture based on an event-driven interaction style employing CEP technology. Each component of the architecture and the interdependencies of components with other elements of the architecture are explained in depth. Here, the detection of the (critical) manufacturing process states (i.e., situations) and the deduction of suitable (re-) actions are emphasized.

The (prototypical) implementation of the envisioned event-driven framework is outlined in Chap. 5. The employed implementation framework, CEP engine, and protocols are delineated. The framework is tested/evaluated in an innovative foundry employing highly-automated foundry machinery. Several industrial scenarios are described, and detection of (critical) process situations and deduction of appropriate (re-) actions are validated. Further, the qualitative and quantitative advantages of the presented methodology are summarized. The research work is summarized and future work is discussed in Chap. 6.

Chapter 2
Problem Description and Fundamentals

In the following chapter, the RTE is introduced as an organizational concept for manufacturing enterprises. The presentation of the RTE starts with a broader discussion of RTE's motivations, fundamentals and principles. Further, the emphasis of presented research and status of development in the realm of RTE are presented. Here, the realization of the RTE in manufacturing can be identified as requiring further attention in ISR.

The event-based processing of information, vertical integration of a manufacturing enterprise and real-time alignment of planned and actual manufacturing process execution are recognized as primary requirements for the realization of an RTE. MES is introduced as a contribution of the engineering community towards the implementation of RTE in manufacturing. In addition, CEP is unveiled as an IT building block for the RTE. However, the liaison between MES and CEP requires further attention in research, which is elaborated in this research work.

2.1 Real-Time Enterprise - Organizational Concept for Manufacturing Enterprises

The RTE is a vision to manage an enterprise in (near) real-time. Consequently, its basic principles are integration of (IT) systems across enterprise borders, automation of value creation processes, and realization of capabilities to satisfy individualized customer demands. Even though RTE is not designated for dedicated industrial sectors, methodologies and implementations of RTE in manufacturing have been widely neglected. In the context of manufacturing enterprises, the implementation of an RTE requires the vertical integration of an enterprise and the realization of closed-loop controls in (near) real-time.

2.1.1 Vision, Principles, and Definition

"Time is money" is a proverb whose relevance has actually increased in recent years. The environment of today's enterprises can be described as dynamic, volatile, and driven by uncertainties [98]. These environmental changes have been significantly pushed by the advent of new (information) *technologies*, especially, revolving around the Internet [4]. Also Gartner, who coined the term RTE [60], is seeing "the effect of inexpensive, powerful telecommunications" [212, 2] and computing power, like Web, e-mail, and so forth, as main drivers for a "shrinking globe" [212, 2]. Besides other factors, these technological developments have affected enterprises' external environments (i.e., demand and competition) as well as internal (organizational) structures and processes.

Information about products and prices is conveniently accessible via the Internet for (potential) customers, who in turn tend to be less loyal to a certain enterprise [4]. Customer demands have fluctuated more, and at the same time, requested products have to adhere to customers' *individual* requirements (i.e., mass customization). Actually, the known transition from a seller's market to a buyer's market has been intensified [165]. Besides the above mentioned technological progress and altered demand behavior, the *globalization* has led to a boost in the number of competitors, especially, from the developing world. Also, because of improvements in communications and logistics, products can be manufactured and distributed all over the globe. Sometimes, this situation is phrased as "design anywhere, build anywhere, sell anywhere" (cf. [167]). Enterprises are no longer operating as part of a flat and linear value chain [207]; rather, they are connected in a complex and multidirectional *value grid* [119].

These tremendous changes in the global economy don't seem to represent just a temporary phase [212]. The vicissitudes of markets, customers' fluctuating behavior, and technological changes are not intermittent, but have to be considered as continual [22]. Unfortunately, the prevalent managerial decision making processes are just supported with historical data, which are often outdated [30]. Besides quality and capability for innovation, time has become a critical success factor and a selling proposition for enterprises [4, 95].

The vision of an RTE addresses the necessity to cope with *time* in a comprehensive way. Gartner's research has initially defined RTE as "an enterprise that competes by using up-to-date-information to progressively remove delays to the management and execution of its critical business processes" [60, 1]. Therein, it is explicitly emphasized that the usage of *up-to-date-information* is required to sustain competitive advantages of an enterprise. Also, Drobik et al. express the strategic character of the RTE as an on-going endeavor that cannot be completed in a fixed, predefined time frame [60]. In the same direction, Abolhassan states the fundamental and comprehensive approach of the RTE [4]. Hence, the RTE is concerned with internal and external processes [12] encompassing SCM (cf. [11, 133]) and CRM (cf. [11, 132]).

Although the work of Alt and Österle pertaining to RTE is mainly focused on extended enterprises, which integrate information across the enterprise's edges, and are consequently denominated as real-time businesses (RTB) [12], it also describes the necessity to coalesce the physical world into the information world [78]. They enunciate the principles of the RTE (or RTB) with the following rules [78]:

- Information created at the point-of-creation (POC) is *immediately*[1] accessible at the point-of-action (POA) for decision making. This requires the realization of *seamless information flows in (near) real-time*. Applying publish-subscribe communication patterns seems to be promising to realize such information flows.
- Information has to be *acquired automatically* at the POC from IT systems (e.g., ERP), programmable logic controllers (PLC) (cf. [138, 141]), radio frequency identification (RFID) chips (cf. [76]), and the like.
- Real-time systems avoid *information buffers*, which are often a result of an inadequate integration of sub-systems.
- There should be no breaks in *semantics* crossing various (IT) systems. Hence, the consideration of standards regarding data semantics is recommended (e.g., ontologies).
- A real-time system has to *select relevant information* that should be made accessible at the POA. Acquired data at the POC has to be filtered, aggregated, and delivered in an adequate timeframe (e.g., seconds, minutes, hours). Thus, only a *significant* subset of all state variables has to be made available for control purposes [148]. Also, Hugos argues that the RTE is "an organization in which specific and relevant information–not just an indiscriminant flood of data–flows continuously to individual people throughout the organization" [120, 1].
- The *decision making* in RTE can be performed at the POA without human interactions or at least with minimum human assistance[2].

Assuming that decision making aims at accomplishing certain objectives, the above principles of an RTE can be summarized to establish (multiple) feedback controls within and across an enterprise [95]. The application of cybernetics and control theory to information systems engineering has been introduced and taught in ISR for a long time (cf. [75]). Hugos explicitly elaborates the theoretical foundations and principles of the RTE as (i) Shannon's information theory [233]; (ii) Wiener's cybernetics [269]; and (iii) Bertalanffy's and Ashby's general system theory [21, 120]. Also, Meyer describes the RTE as an organization that integrates sense-and-respond and learn-and-adapt into its processes [183].

Blood mentions that RTE models often focus on the rapid response to events, but lack considering early warnings as an integral part of the RTE [26]. Motivated by a similar re-thinking, Gartner has adapted his definition of the RTE:

Definition 2.1. (Real-Time Enterprise): The RTE monitors, captures, and analyzes root-cause and overt events that are critical to its success the instant those events occur, to identify new opportunities, avoid mishaps, and minimize delays in core

---

[1] A comprehensive discussion about what is meant by 'immediately' (e.g., soft and hard real-time) is given in Sect. 3.4.

[2] Nevertheless, it is also mentioned in the literature that the incorporation of humans is key to the RTE (cf. [23]).

business processes. The RTE will then exploit that information to progressively remove delays in the management and execution of its critical business processes [176].

This definition accentuates the processing of events for the sake of detection of threats and opportunities. Thus, the (feedback) control within an RTE is preferably *proactive* rather than *reactive*.

2.1.2 Research Scopes, Gaps, and Challenges

By definition, the RTE is entirely influencing an enterprise [4], i.e., its strategies, organizational structures, internal processes, and external interactions with market participants. Nevertheless, ISR revolving around the design of the RTE has concentrated on the extended enterprise. The need to react to changes occurring in the *external environment* of an enterprise is often expressed with vigor (cf. [12, 225]). For instance, Alt and Österle argue that enterprise integration (EI) has focused on internal processes whereas external processes with suppliers and customers have been neglected [12]. Furthermore, they state the aim of ERP systems, like SAP® R/3, is the real-time dissemination of information across humans, organizational units, and even enterprises.

Although it is vital and inevitable to *horizontally* integrate an enterprise with its external environment, as mentioned by Alt and Österle, its vertical integration is just as important. The *vertical* enterprise integration across enterprise levels, which is capable of linking planned and actual process execution, can be seen as a prerequisite for the realization of an RTE in manufacturing [95]. In addition to enterprise integration, automation as well as individualization has to be achieved or appropriately addressed in an RTE [78]. The challenges of vertical integration of an (manufacturing) enterprise are summarized in the following paragraphs.

In Fig. 2.1, a simplified model of a manufacturing enterprise is depicted as a triangle composed of several enterprise levels. The enterprise's external environment is delineated as SCM and CRM, and their horizontal integration with the enterprise is denominated in this research work as an extended enterprise[3]. Extensive research has been carried out to automate and integrate business processes at the tactical level of an enterprise. Primarily, administrative, purchasing, and financial processes are executed at this enterprise level. According to standard VDI 5600, this level is called enterprise control level [255], and it is often supported by ERP systems. On the contrary, actual value creation is performed at the resource level or manufacturing level. The intermediary between the enterprise control level and the manufacturing level is the manufacturing control or operational level, which is in charge of providing a comprehensive control of manufacturing processes. The vertical integration of a manufacturing enterprise is challenging as enterprise levels

---

[3] Contrarily, the term extended enterprise can also be understood as the integration of enterprises in a value network or value chain.

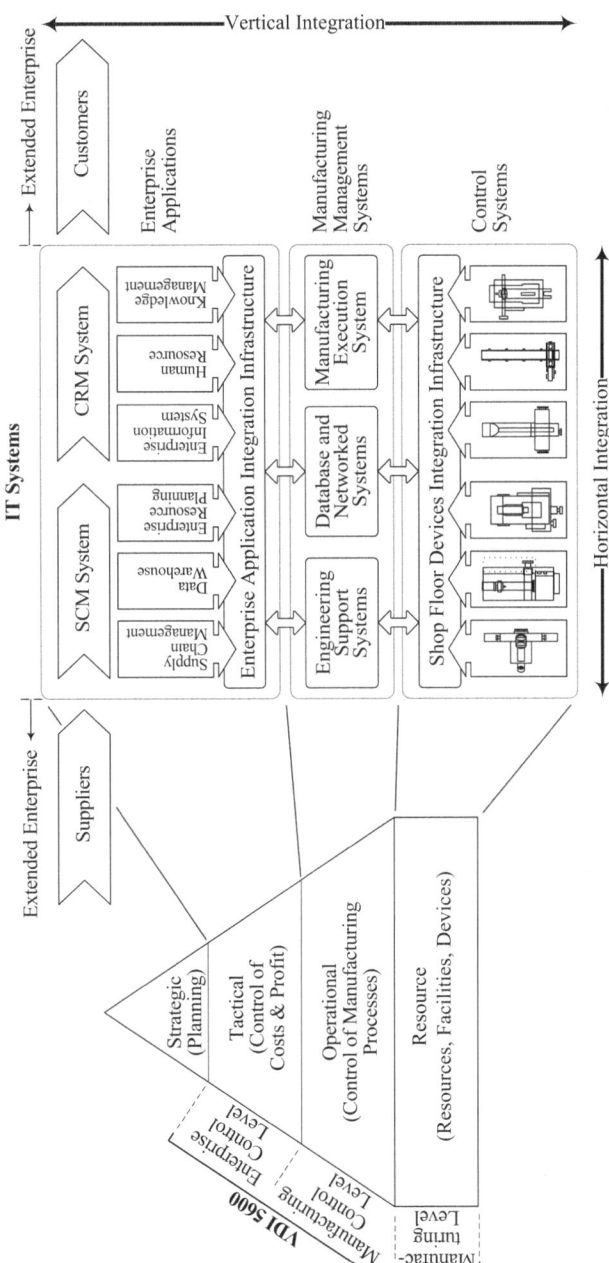

Fig. 2.1 Horizontal and vertical enterprise integration as a prerequisite for the RTE, and some characteristics of IT support at different enterprise levels (adapted from [101, 117]).

are associated with different time horizons varying from (milli-) seconds over hours and days to quarters and years. In other words, business processes located at the tactical level are predominantly executed *offline* whereas manufacturing processes situated at the shop floor level are executed *online*. Briefly, these properties of the enterprise levels result in a semantic and temporal vertical integration gap.

MES are located at the manufacturing control level and try to bridge the vertical integration gap (cf. [255]). However, major problems with respect to MES are reported in the literature [201], primarily, because inflexible and proprietary interfaces between shop floor resources and enterprise applications (e.g., ERP system) impede a seamless (vertical) integration [140]. Therefore, research has been carried out to exploit the SOA paradigm also at the manufacturing level to overcome these shortcomings. Prototypical implementations of SOA-enabled devices (e.g., programmable logic controllers) at the manufacturing level have been examined in EU funded projects (cf. [29, 237]). Nevertheless, further developments are essential to refine and improve these implementations, and to foster their acceptance and dissemination in industry.

In its final configuration, the RTE envisions the integration of feedback controls between POCs and POAs, hence exceeding basic integration of data. Instead, a continual detection and analysis of events, which happen during execution of value creation processes, and a manipulation of these processes by dispatching control commands has to be performed in (near) real-time. The RTE's goal of a comprehensive real-time control of (manufacturing) processes necessitates the examination of (shop floor) events with respect to their administrative and financial context. Logically, the basic vertical integration of enterprise data can be considered as a *necessary condition* of the RTE whereas event processing and (automatic) deduction of appropriate (re-) actions can be described as its *sufficient condition*.

2.1.3 Emergent Enablers

The definition of the RTE (cf. Def. 2.1) does not encompass explicit descriptions of (information) technologies to be employed for its realization [23, 60]. Rather, it is focused on business aspects, where IT is just seen as an enabler for the RTE. According to Fleisch and Österle, the key concepts, which have to be addressed by IT within an RTE, are: (i) integration; (ii) automation; and (iii) individualization [78]. In their solution, the integration of an enterprise has been realized with *EAI solutions*, capable of connecting various enterprise applications, like an ERP system and a product data management (PDM) system, at the enterprise control level. Further, a *portal solution* based on principles of an SOA has been presented to also integrate SCM and CRM effectively [11]. In general, *SOA* is seen as a substantial enabler of RTE as it provides flexible and adaptable infrastructures [11].

In addition, Schulte discusses the event-driven behavior of the RTE and the need for the application of events, messaging, and publish and subscribe concepts at the business level [230]. EDA has been mentioned as another enabler for the RTE

[95]. Moreover, the automatic detection of opportunities and threats, which are inherent parts of the RTE vision, are notably more than a crude analysis of simple events. Simple events originated by manufacturing processes have to be related, e.g., in temporal and causal terms. Hence, CEP is currently mentioned as another technological building block for the RTE (cf. [41, 164]).

2.1.4 Summary and Research Goals

Based on the above mentioned explanations, the current situation with regard to the realization of the RTE in manufacturing can be summarized as follows:

- RTE is a comprehensive concept covering all aspects of an (manufacturing) enterprise. However, research has primarily focused on the topmost enterprise levels and the extended enterprise, but has neglected the vertical integration of an enterprise with its value creation processes at the manufacturing level.
- Available MES solutions address the vertical integration and are indeed considered as enablers for the RTE. Nevertheless, MES still possesses some shortcomings in the interface with shop floor resources [118].
- Further, promising event-driven IT, like EDA and CEP, have been introduced (into the RTE) for administrative and financial enterprise processes. However, a concise framework is missing to link MES and CEP. The application of CEP in manufacturing is currently limited and requires further consideration.

The consequences of the aforementioned situation within manufacturing enterprises are numerous: (i) the *restricted transparency* of value creation processes impedes accurate, consistent and timely decision making processes; (ii) critical (complex) situations cannot be *sensed* in (near) real-time, and hence, necessary *responses* are induced relatively late to overcome or mitigate these situations; and (iii) enterprise-wide *learning* is hindered, and further, *adaptation* is not performed on the basis of experienced facts (empiricism) about the processes' outcome (i.e., quality, performance, maintenance, inventory).

This research work presents a framework comprising a process model and software architecture, which both contribute to the realization of the RTE in manufacturing. The IT part of the framework is based on principles of EDA and CEP, enforcing the vertical integration of enterprise levels and facilitating the introduction of comprehensive feedback controls in (near) real-time. This contributes to the implementation of the RTE principles (i.e., *learn-and-adapt* and *sense-and-respond*) within a manufacturing enterprise. The results presented in this research work have been achieved and experienced during a three-year industrial project with the Ohm & Häner Metallwerk GmbH & Co. KG, Olpe, Germany. This enterprise manufactures high quality aluminum sand castings employing state-of-the-art machinery and a high degree of automation. In addition, further results could be attained during an 18 month research project funded by the Federal Ministry of Economics and Technology (BMWi), Germany.

2.2 Principles and Fundamentals of Event Processing

As the envisioned IT framework is grounded on the EDA paradigm and exploits CEP technology, important principles are outlined with regard to the same. These descriptions are mainly centered on CEP as an innovative enabler for sophisticated monitoring and control of value creation processes. In addition, EDA and CEP are discussed in contrast to their predecessors, and competing and complementary concepts.

2.2.1 Events and their Processing as a Fundamental Requirement for Business

Although events are an inherent part of the world, they have not been widely used in business [229]. Oftentimes, one has just dealt with simple events, i.e., has focused on low-level events, like in network monitoring [161]. Higher-level events that can be exploited for monitoring and control of business processes have been neglected, as they have not been supported by IT in a systematic way. The correlation among business pressures, management strategies, information system requirements, and event processing has been delineated by Chandy and Schulte, as depicted in Fig. 2.2 [41]. The demand for a systematic and computerized processing of events is clearly illustrated in the figure.

Obviously, the business pressures already mentioned in previous paragraphs (and listed in Fig. 2.2) have intensified the need for novel management strategies, like RTE. Besides RTE, the agile, adaptive, predictive, zero-latency and event-driven enterprise can also be referred to as having innovative strategies; note that zero-latency and event-driven enterprise are synonymous with RTE [230]. Putting enterprise strategies into practice, one often has to pursue the following simplified steps: (i) recognition and anticipation of threats and opportunities for the enterprise; (ii) development of strategies concerning business, products, and technologies; (iii) (re-) design of the organizational structures and business processes in accordance with the selected strategies (i.e., structure follows strategy); and (iv) implementation of the processes, probably by employing appropriate IT [87]. Hence, RTE and similar management strategies result in serious consequences for the enterprise's processes and systems. In the case of RTE, they can be summarized as timeliness, agility, and information availability [41]. In technical systems, *timeliness* is often associated with the term (low) *latency*. The meaning of latency depends on the context wherein it is being used. In computer networks, latency is the time required for a data packet to get from a sender to a receiver, which is influenced by, e.g., the packet size, communication medium, and routing [25]. More generally, latency is understood as response time in computer systems, i.e., the time between a request to a system and its response. Similarly, in real-time control systems, delays are described for controlled objects (e.g., a heater) and for controlling computer sys-

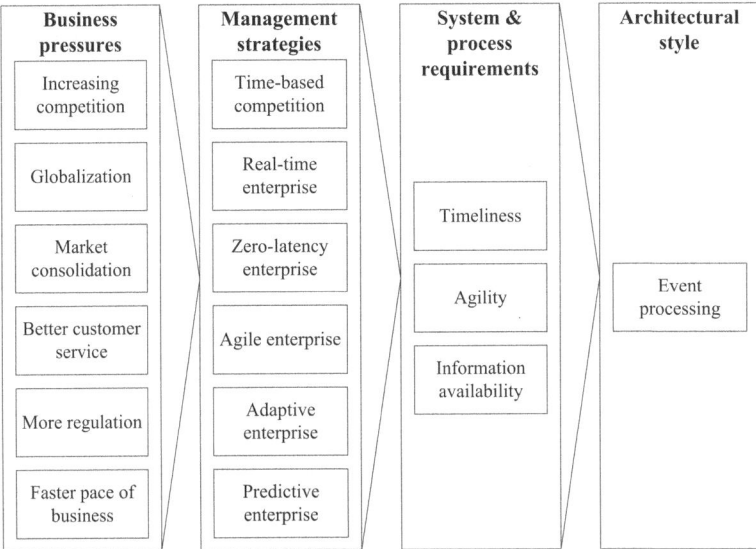

Fig. 2.2 Correlation among business pressures, management strategies, system & process requirements, and event processing (adapted from [41]).

tems (e.g., programmable logic controller). For the latter, the delay (i.e., latency) describes the time between the observation of the controlled object (e.g., measured temperature of the heater) and the push of a derived output (i.e., reaction) from the controlling computer system to an actuator (e.g., thermostat) [148]. Finally, latency can also be defined with regard to event processing systems, where latency is the difference between an event's egress time and its stimulus time (i.e., the time an event arrives at an event processing system) [40]. In a business context, low latency can be considered as the execution of single activities or even orchestrated processes with celerity [41]. In addition, it also connotes timely decision making processes and immediate reactions to events.

Because of turbulences in markets, technological progress, and the like, enterprises strive for flexibility and agility. Unfortunately, there is no consensus about the definition of flexibility and agility [92]. *Flexibility* has been addressed in business process management (BPM), where "process flexibility can be seen as the ability to deal with both foreseen and unforeseen changes, by varying or adapting those parts of the business process that are affected by them, whilst retaining the essential format of those parts that are not impacted by the variations" [227, 17]. Also, flexibility has been considered for manufacturing processes as "the ability of a system to change its behavior without changing its configuration" [267]. In addition, Wadhwa and Rao consider flexibility as a predetermined response to predictable changes, and further, describe the scope of flexibility on single systems [262].

The concept of an agile enterprise was first worked out in the 1990s [57]. Later, a significant number of definitions of *agility* were based on flexibility. Oftentimes, agility is described as flexibility complemented with quickness and speed (cf. [262]). Oosterhout et al. analyze literature pertaining to agility, and conclude that "business agility is the ability to sense highly uncertain external and internal changes, and respond to them reactively or proactively, based on innovation of the internal operational processes, involving the customer in exploration and exploitation activities, while leveraging the capabilities of partners in the business network" [197, 53-54]. Intriguingly, this definition of *business agility* encounters the main aspects of the RTE (cf. Sect. 2.1): sense-and-response to uncertain changes, response in a reactive or even proactive manner, inclusion of processes related to customers, and encompassing partners in the supply chain.

After an investigation of the discourses on flexibility and agility across various research disciplines, agility seems to describe a higher form of changeability. Accordingly, a maturity model for manufacturing system changeability has been described, starting from changeover ability via reconfigurability, flexibility, and transformability to agility [268]. In contrast to flexibility, agility's scope is on groups of systems, its focus is on responsiveness (i.e., quickness), and it must cope with uncertain and unknown changes, must respond proactively, and control (processes) dynamically [262]. As enterprises and their processes are largely reliant on IT, the discussion on (organizational) agility has also affected the information system's agility, which can be differentiated into technical infrastructure agility, information systems process agility, and human characteristics [173].

SOA has a high potential to substantially contribute to a company's agility [8]. This potential can be derived from SOA's inherent concept of loosely coupled software components, existing standards for implementing SOA-based components, and realization of enterprise interoperability build on top of SOA [202]. Following SOA, EDA has been discussed as a paradigm that leverages events to achieve even "minimally coupled" software components [41, 34]. Events are emitted by event producers (cf. POC in Sect. 2.1.1) and received by multiple event consumers (cf. POA in Sect. 2.1.1), which are interested in the emitted events. The event-driven communication within EDA implies that the software components do not have any preconception about each other. Hence, EDA augments loose coupling and fosters information system agility. However, EDA does not need to be perceived as contradicting SOA, rather as complementing SOA to achieve business agility. Recently, there is a trend to enhance EDA by employing CEP, i.e., analyzing the (causal and temporal) relationships of events [46].

In addition to timeliness and agility, the *availability of information* is considered as an indispensable requirement to attain the vision of the RTE. Voluminous amounts of information have been made available to users in the era of the Internet. Search engines, like Google^{TM}, support users searching for relevant information. Similarly, *enterprise search* supports enterprise members to retrieve appropriate documents, like 2D drawings and 3D models of components (products), product calculations, and process plans [100]. Among others, the retrieved information considerably assists knowledge-intensive processes (e.g., offer process) and improves the accuracy and consistency of decision making processes. The integration of various information systems and corresponding data sources is a prerequisite for the comprehensive availability of information. In this direction, data warehousing and master data management aim to achieve data consistency [41], and further, form a basis for aggregating information with higher significance (e.g., key performance indicators (KPIs)) for decision makers. In addition, the transparency of processes can be enhanced by employing such information systems.

Nevertheless, the above approaches to increase the information availability primarily follow a request-reply communication pattern (cf. Sect. 2.2.5). In essence, available information has to be *pulled* out of the aforementioned systems by either an information system or enterprise member that has a need for using that information. In contrast to this offline information propagation, information can also be disseminated in a more instant fashion. Whenever new information has been created, it will be *pushed* immediately to relevant decision makers (i.e., human or information system acting on behalf of a human). The event-driven interaction of (software) components of EDA and CEP adheres to this online/real-time delivery of information.

Emergent technologies, like event processing based on EDA and CEP, substantially contribute to the realization of the RTE. The inherent characteristics of EDA and CEP, like minimal coupling and flexibility, strongly address the main system and process requirements arising from the RTE vision (i.e., latency, agility, and availability of up-to-date information). Nevertheless, it has to be clarified that time- and request-driven process behavior and its corresponding support by IT will not become obsolete with the introduction of event-driven architectural styles. Rather, in many instances, software architectures are capitalized on various complementary interaction styles. After motivating the employment of EDA and CEP in the context of the RTE, the subsequent paragraphs will delve into fundamentals, principles, and techniques of EDA and CEP.

2.2.2 Definitions and Circumscriptions of Events

An elaboration of basics revolving around event processing and corresponding architectural styles should start with a definition of the term *event*. The common-sense understanding of what an event is can be found in nearly every dictionary. For instance, Merriam Webster Learner's Dictionary states that an event is "something (especially something important or notable) that happens" [180]. Relevance and importance of an event are accentuated in the dictionary's definition. Accordingly, events, which are noteworthy with regard to business processes, are described as *business events*. They are "meaningful to conducting commercial, industrial, governmental, or trade activities" [41, 1]. Also, a business event can be described as "anything that *changes the status* of the business in some way" [5].

The importance of a business event largely depends on its semantics and situational context. For instance, a business event can simply signal the arrival of an email but also something more important, like the cancellation of a customer's order. The time when an event is raised can be considered as an outstanding part of its situational context [164]. In most instances, for example, a machine's malfunction is important at the time of its occurrence but probably not after years have passed. Similarly, this issue is discussed in the context of real-time systems as the *accuracy interval* of a real-time image, which represents an observation of a real-time entity (i.e., significant state variable of a controlled object, like a car or plant) at a certain point in (physical) time [148]. The observation of the position of a robot in a plant, for example, is only temporally accurate.

Chandy and Schulte clarify that (business) events cannot be entirely foreseen [41], i.e., events happen spontaneously and cannot be fully anticipated. For instance, a production resource can break down at any time, thus causing serious delays in fulfilling customer orders. Further, Luckham summarizes the indispensability to make use of events in business [164]. The question is not *if* an enterprise should use events, rather *how* to use events in an effective and efficient way. Nonetheless, IT systems have handled events in a certain manner [41].

Etzion and Niblett consider an event as "an occurrence within a particular system or domain; it is something that has happened, or is contemplated as having happened in the domain" [67, 4]. Noteworthy, their definition stresses the fact that an event doesn't necessarily need to correspond to a real world happening. The indication of disturbances in a manufacturing process may be suspected by analyzing multiple event occurrences, which individually considered would not reveal any malfunction. Also, not every event is directly *observable*, either because of missing event producers (e.g., sensors) or a higher abstraction of the event. Similarly, Luckham introduces the distinction between *actual events* and *virtual events*, where the latter is "an event that is thought of as happening" [164, 29]. Also, the absence of an event can be meaningful within a business process (e.g., an intended action has not taken place). This kind of events can be designated as *non-events* [67].

In contrast to these broad event definitions, events, which are exploited in the IT realm, are defined as programming entities applicable to represent real world event occurrences [67]. In the context of the processing of events within computer systems, events are represented as *event objects* [41, 67, 164]. These objects are records of an activity in a system or domain, and possess three aspects [161]:

- An event's *form* is an object encompassing attributes and data components. Although the event form is defined as an object with regard to computerized processing, it is not necessarily considered as an object from object-oriented programming (OOP) paradigm [41, 67].
- The activity signified by an event is called the event's *significance*. Event object's attributes are used to describe the activity.
- The activity, which is indicated by an event, often has relations to other activities. These *relativities* are, for instance, defined based on time, causality, and aggregation. Consequently, events that signify certain activities have various relationships with other events.

The latter aspect of an event object is of paramount importance as it clearly separates events from messages. Events do not purely carry descriptive information about an activity; rather, events also express temporal and causal relationships to other events [164]. Nevertheless, messages can be employed as a mechanism to convey events, thus are sometimes called notifications [67].

In addition to the aforementioned *happening view* on events, an event can also be interpreted as a report of a state change [41]. This explanation of an event is often used in IT realms, where it has been described as the change of a data value [243]. Contrarily, Faison defines an event as "a detectable condition that can trigger a notification" [69, 71]. Here, a certain number of logically connected predicates constitute a condition, which can be detected and, in turn, triggers a notification indicating that the condition has been fulfilled. The happening, state change, and condition detected view of an event are overlapping [41] and are focused on different aspects of an event. Although it is vital to keep all elaborated aspects of an event in mind, the definition of the event processing technical society (EPTS) is preferred in the context of this research work as it is the most widely referenced definition in the literature surrounding event processing:

Definition 2.2. (Event): Anything that happens, or is contemplated as happening [163, 5].

Accordingly, an event object, subjected to a computerized processing of events, is defined as follows:

Definition 2.3. (Event Object): An object that represents, encodes, or records an event, generally for the purpose of computer processing [163, 5].

Thus far, the transformation of a (real world) event to a corresponding event object has not been standardized [67, 164]. It is clear that an event object doesn't have to include all possible attributes of an event, which has been observed. Similarly in real-time systems, only a subset of state variables out of all state variables of a controlled object (e.g., robot) has to be gathered that is significant for the intended purpose (e.g., control of the robot) [148].

An event standard defined by OASIS is called the WSDM Event Format (WEF) that encompasses an event's identification, information about the event source, details about the event sink, a message with regard to what has happened, and timestamps [37]. The format is based on the eXtensible Modeling Language (XML) as is common with other Internet related standards. The intended use of the WEF is in the field of systems management applications and network management [67]. Further, the Service Interface for Real Time Information (SIRI) is a standard for the exchange of real time information in public transportation [235]. Finally, Taylor et al. suggest leveraging simple object access protocol (SOAP) messages, which are widely used in current SOA implementations, as a foundation for an EDA [243].

In the case of a computerized processing of events, it is noteworthy to perform the following steps in accordance with this event classification, as depicted in Fig. 2.3: (i) identify events, which are of (high) relevance for an aspired purpose (e.g., control of a process); (ii) identified business events or other events have to be made observable, either direct as detectable conditions or indirect as abstract events; and (iii) observable events have to be codified as event objects. In the context of this research work only business events are considered.

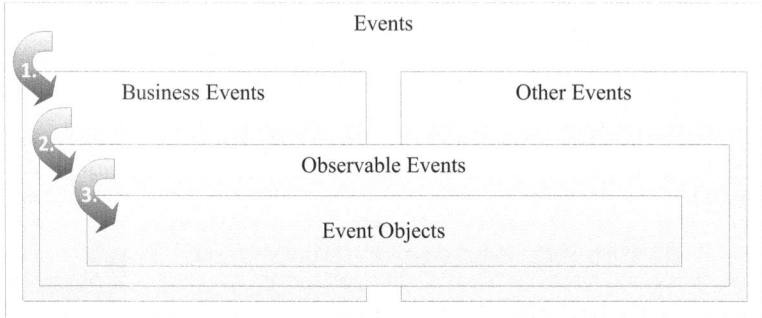

Fig. 2.3 Classification of various event definitions, and the path from real world events to event objects (adapted from [41]).

2.2.3 Event Clouds, Event Streams, and Complex Events

In the pioneering book "The Power of Events" written by David Luckham (cf. [161]) about event processing, the situation of enterprises is described as being involved with a *cloud of events* (cf. Fig. 2.4). Events are produced continually by a plethora of (information) systems both inside and outside of an enterprise. In the case of automated manufacturing processes, huge amounts of data are produced by control systems, for instance, by PLCs. Also, stronger collaboration with suppliers and partners in value creation networks implies increasing the exchange of information (and events) across enterprise borders. In the following discussion, it is assumed that the events within the event cloud are observable and a corresponding event object exists for each event occurrence. The event objects in the event cloud are of various event types. An event type is a "specification for a set of event objects that have the same semantic intent and the same structure; every event object is considered to be an instance of an event type" [67, 62]. Further, it is often unknown or uncertain when an event in the event cloud emerges. Also, the events in the event cloud are not necessarily *linearly* ordered by time. Consequently, the EPTS has defined the event cloud as follows:

Definition 2.4. (Event Cloud): A partially ordered set of events (poset), either bounded or unbounded, where the partial orderings are imposed by the causal, timing, and other relationships between events [163, 14].

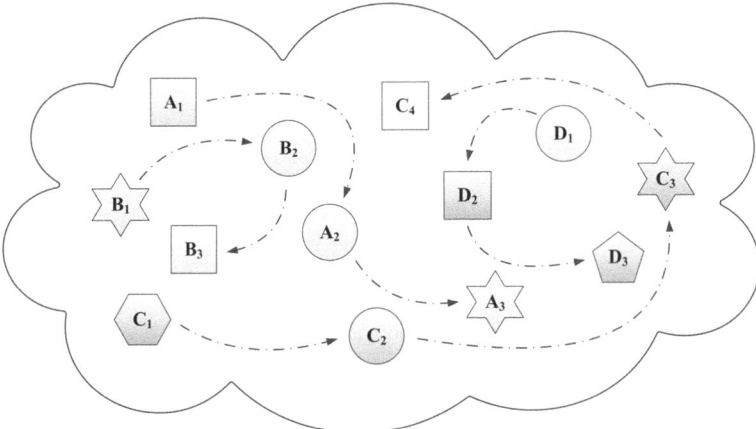

Fig. 2.4 Illustration of the event cloud composed of various event instances, and outline of different relations between events (adapted from [164]).

In contrast, an event stream is "a linearly ordered sequence of events" [163, 13]. In most instances, the incoming events are linearly ordered according to their *arrival time*, i.e., the time an event object is observed by a system, which is considered as an important criterion. Therefore, an event stream can be a part of the event cloud, or it can be considered as a special event cloud with linearly ordered events [162].

The above mentioned differentiation between event cloud and event stream is of paramount importance as it clearly separates CEP from the plain processing on event streams, i.e., event stream processing (ESP). The latter focuses on high-speed querying of data in event streams, and considers events in the order of their arrival time, which is fast and consumes less memory [162]. Contrarily, CEP incorporates more sophisticated relations between events, thus deals with more advanced scenarios and more complex business processes.

The events in the event cloud have various relationships. The most prominent are time, causality, and aggregation. Simplifying, event relationships can be employed to form a complex event based on other (base or simple) events. Consequently, the EPTS has defined a complex event as follows:

Definition 2.5. (Complex Event): An event that summarizes, represents, or denotes a set of other events [163, 8].

In this context, an *event pattern* is a template that contains other event templates, relational operators, and variables [163], and further, can be used to describe a complex event. In Fig. 2.5, a semi-formal symbolism of events, event relations, and an event pattern are illustrated. The presented event relationships are most usual in event processing, and have become main building blocks for event patterns. An advanced analysis of events within the event cloud can be performed by investigation of the aforementioned event patterns. Consequently, this analysis leads to a

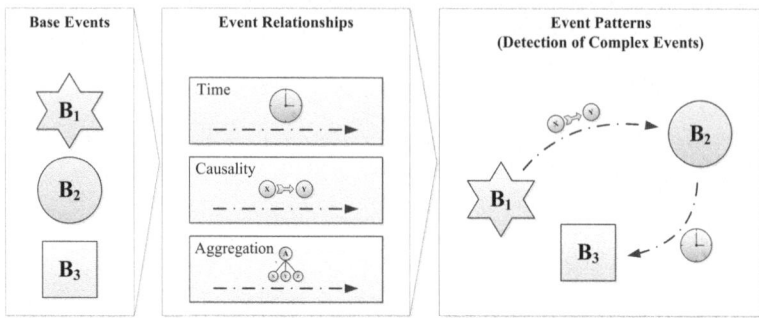

Fig. 2.5 Event relationships are forming complex events employing (base or simple) events.

hierarchical abstraction of events. Hence, event patterns are instruments to structure the event cloud (i.e., define a hierarchy of events). The principle of abstracting
events with event patterns is depicted in Fig. 2.6. At the lowest level–also denominated as abstraction level 0–the event cloud is composed (primarily) of simple or
base events[4]. A simple or base event is "an event that is not viewed as summarizing, representing, or denoting a set of other events" [163, 10]. Within the next
overlying level–named as pattern level 1–event patterns describe complex events at
the abstraction level 1. An event pattern is expressed by event types, variables, and
relational operators. The detection of an event pattern within the event cloud can be
divided in three steps: (i) filtering of the relevant event types; (ii) matching of event
instances that apply to the conditions and relational operators defined in the pattern;
and (iii) derivation of a pattern matching set [67]. The pattern matching set is composed of derived events, which are input for the next abstraction level. The process
of abstracting events is pursued until the topmost abstraction level N is reached.

Event abstraction is defined as "the relationship between a complex event and
the other events that it denotes, summarizes, or otherwise represents" [163, 8]. The
recognition and usage of higher-level events is crucial for managerial purposes (cf.
[41]). In other words, the exclusive use of simple events is not sufficient for the
control of enterprise processes owing to reduced entropy (i.e., information content)
[265].

Further, an increasing abstraction along hierarchical levels concurrently involves
a decreasing number of derived events on higher abstraction levels. These derived
events have higher entropy. As a result, the concept of event abstractions contributes
to the reduction of complexity and increase of relevant information. As such, it
supports an enterprise to cope with the cloud of too much information [18, 164]. In
summary, higher-level events can be formed by applying event patterns on lower-
level events. The temporal, causal, and aggregation relationships for abstracting
events are elaborated in the subsequent paragraphs.

---

[4] Actually, an event cloud is composed of simple and complex events, because derived complex
events can (also) be appended to the event cloud.

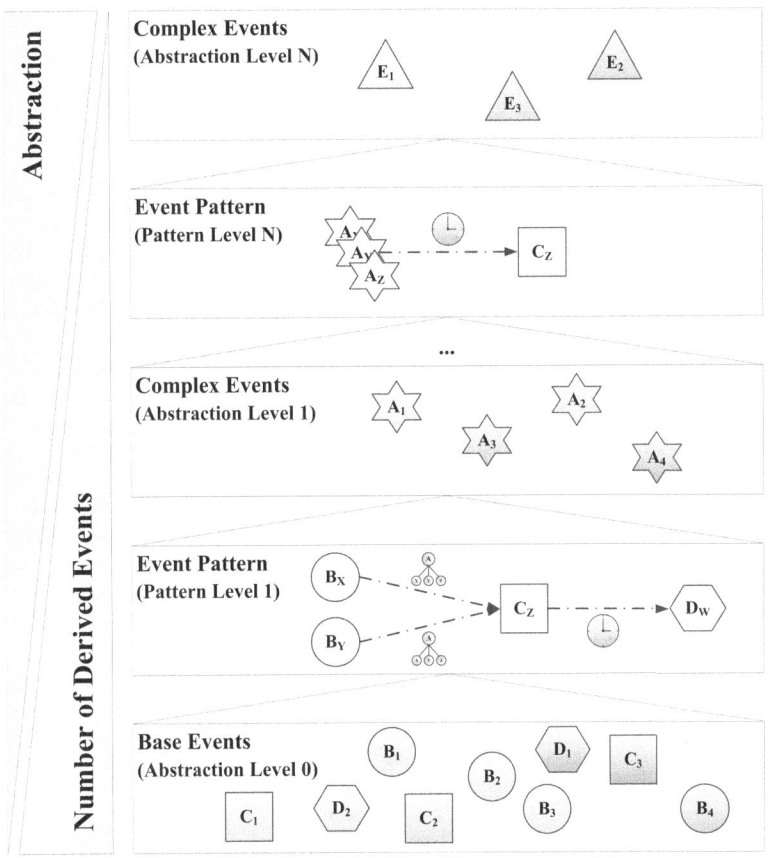

Abstraction

Number of Derived Events

Fig. 2.6 Hierarchical abstraction of events using event patterns (adapted from [163]).

Usually, an event happens at a certain *instant in time*. This instant in time is denoted as the event's *creation time* that can be added as an attribute to the corresponding event object. Similarly, the *occurrence time* records the time at which an event occurred in a system [67], which is also called *application time* [53]. Further, an event gets observed and enters an event processing system—either CEP or ESP—at its *arrival time*, which is also termed *detection time* [67] or *system time* [53][5].

---

[5] In some instances, it is inevitable to consider event occurrence times (i.e., application times) while processing events (in streams). Otherwise, anomalies might occur, which stem from the fact that there is no global clock in a distributed system and non-controllable latencies of event delivery exist. Hence, for instance, read-too-late or write-too-late anomalies are reported in the literature (cf. [53]). However, these anomalies most likely happen in scenarios in which related

In any case, events can be linearly ordered by time–either by their creation or arrival time. In addition, Etzion and Niblett discuss whether an event necessarily happens at an instant in time (i.e., with zero time duration), or rather, possess a start and end time (i.e., time interval with a non-zero time duration) [67]. Similarly, Chandy and Schulte state that events can be interpreted as being instantaneous, or alternatively, events have a measurable duration [41]. The former is the prevailing view in the context of a computerized processing of events in which events are understood as state transitions. However, an event expressed with certain duration can be transformed into two separate events, i.e., one for representing the start and one for marking the end. The other way around, two simple events, signifying start and end events, can constitute a complex event (i.e., interval event) encompassing these events as members.

In Fig. 2.7, thirteen possible temporal relations between two events X and Y are depicted[6]. Each event is an instance of a certain event type wherein t_s and t_e are the start time and end time, respectively [135]. Consequently, a temporal relation of two events can be formalized defining the relations between their start and end times. For any temporal relation, which is formulated in Fig. 2.7, an inverse relation is articulated, a pictorial illustration is drawn, and endpoint constraints are listed. Causality can be commonly defined as "the relationship between something that happens or exists and the thing that causes it" [179]. In the context of event processing, causality is strongly related to activities, which are *dependent* on each other [161]. An event B is caused by another event A, if event A had to happen in order for B to happen [163]. Apart from this computational definition, causality has been discussed extensively in philosophy and statistics [161, 163]. In the latter, causality is expressed as probability or likelihood. However, it is a misconception to consider causality in a computerized event processing system as a *cause-and-effect relationship* in an ordinary way; in fact, it implies that a causal event is indispensable although not sufficient to determine that the subsequent event will occur [41].

Causality tracking can be used to determine possible causal relationships between events, which is a crucial problem in conjunction with distributed computing systems [112]. These events either signify send events, receipt events, or internal events [172]. However in that case, causality can be reduced to a "happened before" relation, and thus, is dependent on arrival times (i.e., timestamps within the events) [177]. The happened before relation provides a partial ordering of events in a distributed system [154], and as such, can be exploited for an illustration of causal event relationships in the form of directed acyclic graphs (DAG) [211]. The reliance on event timestamps implies that the tracked causality *potentially* exists (i.e., *potential causality*) [172], which is different from the ordinary understanding of causality. The antagonisms to causally-related events are independent events, which are considered in distributed computing systems as to happen concurrently [112].

---

events occur within a relatively narrow span of time (e.g., machine control that requires extremely low latencies).

[6] Originally, Allen outlined a theory of action and time in the field of artificial intelligence [10]. Allen's taxonomy of temporal relationships is adopted by Kam and Fu to describe the temporal relationships among events [135].

Relation	Relation for Inverse	Pictorial Example	Endpoint Constraints
X *before* Y	Y *after* X		$X.t_e < Y.t_s$
X *equal* Y (X *coincides* Y)	Y *equal* X (Y *coincides* X)	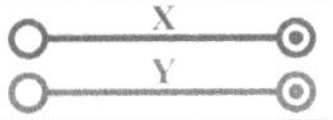	$X.t_s = Y.t_s$ $X.t_e = Y.t_e$
X *meets* Y	Y *met-by* X	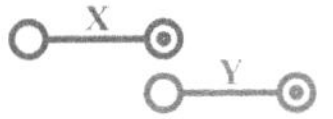	$X.t_e = Y.t_s$
X *overlaps* Y	Y *overlapped-by* X	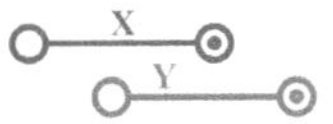	$X.t_s < Y.t_s$ $X.t_e > Y.t_s$ $X.t_e < Y.t_e$
X *during* Y	Y *contains* X	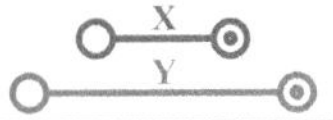	$X.t_s > Y.t_s$ $X.t_e < Y.t_e$
X *starts* Y	Y *started-by* X	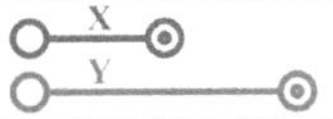	$X.t_s = Y.t_s$ $X.t_e < Y.t_e$
X *finishes* Y	Y *finished-by* X	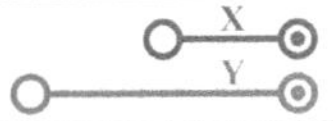	$X.t_s > Y.t_s$ $X.t_e = Y.t_e$

Fig. 2.7 Temporal relationships between event (objects) X and Y adapted from [10, 135]. Each (complex) event has a start time denoted as t_s, and an end time denoted as t_e.

This discussion on causality in distributed computing systems reveals a close relation between time and causality. This relation has been expressed in a *cause-time axiom*: "If event A caused event B in system S, then no clock in S gives B an earlier timestamp than it gives A" [161, 96]. This obvious law of nature regarding causality has been exploited by Lamport to establish logical clocks (i.e., for the previously mentioned partial ordering of events) [90, 154, 242].

Apart from causality in distributed systems, *causal modeling* can be performed to uncover causality between events. The discovery of event dependencies consists of three steps: (i) specifying the event sets to be monitored; (ii) investigation of event causality by interviewing enterprise members, analyzing (IT) system documentation, and/or examining event logs[7]; and (iii) construction of a formal causal model encompassing rules for causality tracking (detection) [161]. However, although some notions for causality exist (e.g., RAPIDE system has provided a syntax for (in-) dependency between events), Luckham argues that a more explicit support for causality is a future step of event processing [164].

Finally, the definition of the term complex event (cf. Def. 2.5) already suggested a combination of events. Therefore, Etzion and Niblett broadly discuss event aggregation and composition as *stateful* event transformations [67]. Apart from a simple combination of events to an aggregated event, *derivation functions* can also be ap-

---

[7] A systematic analysis of event logs can be performed, for instance, with process mining methods (cf. [1]).

plied to event attribute values extracted from a set of events, and further, are used to form an aggregated event [67]. In essence, an aggregated event summarizes information contained in underlying events, and as such, represents information that cannot be directly observed. Many derivation functions have been adopted from (extended) relational algebra, which is commonly used with relational databases. Examples of these aggregation operations are minimum, maximum, average, count, and sum. Other derivation functions are inherited from statistics including standard deviation, median, and so forth.

Summarizing, events are generated by a multitude of systems, especially, along the execution of value creation processes. Moreover, events can be considered as drivers for value creation processes [161]. These explanations have shed light on various relationships between events (i.e., temporal, causal, aggregation). Further, the consideration of event relationships leads to an abstraction of (complex) events, and thereby, provides a hierarchical organization of events in the event cloud. Luckham has characterized the situation concerning value creation processes and the resulting challenges and demands with regard to their (IT) support as follows [161]. Firstly, in contrast to the common assumption of linearly structured processes supported with workflow engines, future processes will possess *nonlinear* sequences of activities to a greater extent. Hence, transitions between process steps can be initiated only after an advanced analysis of *sets of complex events*. Secondly, processes will become even more *agile* (cf. the discussion in Sect. 2.2.1), which demands a higher degree of IT system *flexibility*. Finally, *exceptions* during process execution will happen more often, thus require capabilities for real-time (process) *awareness* and incorporation of event causalities.

2.2.4 Historical Background pertaining to Processing of Events

The processing and usage of events in IT has a longer tradition than today's CEP technologies. Before delving into the details of state-of-the-art event processing based on EDA and CEP, a compact historical outline of developments surrounding events is presented, as depicted in Fig. 2.8. Further, this description of event (processing) history is *not complete*; it highlights some crucial developments in the context of the usage of events in various IT domains. However, the temporal delimitation of the presented technological achievements cannot (always) be defined with absolute accuracy, and thus, this briefing cannot claim to be the only possible interpretation of history. Although the developments are limned in isolation, they are actually often interdependent and some developments could also be assigned to another IT domain. The presented developments are predecessors of modern EDA and CEP solutions, and thus, it is vital to explore their motivations, backgrounds, and technical foundations. Moreover, the consideration of the various roots of CEP is reasonable as CEP has been only recognized in very recent years as an independent field [61]. In fact, EDA and CEP exploit developments from many of the technolo-

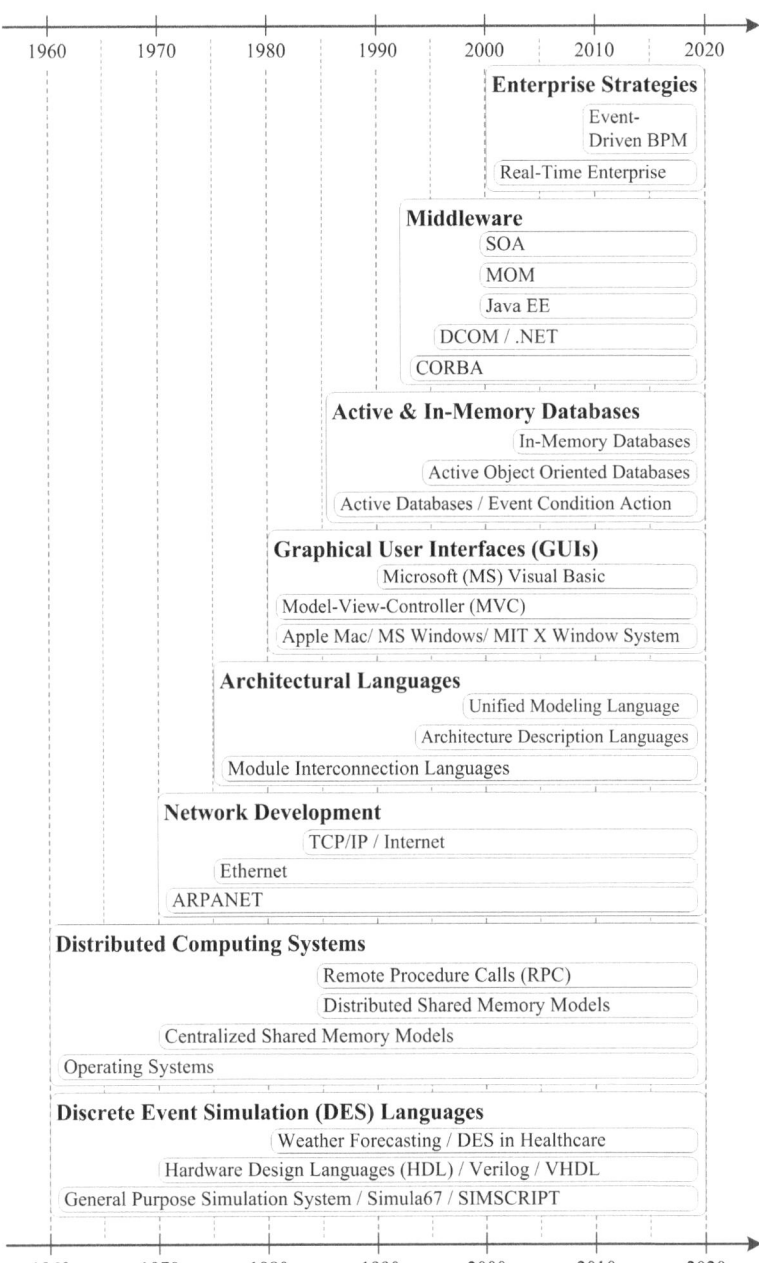

Fig. 2.8 History of events and predecessors contributed to event-driven architectures and (complex) event processing (primarily merged from [69, 161]).

gies listed in Fig. 2.8. The following descriptions are considerably inspired by the work of Luckham and Faison (cf. [69, 161]).

Discrete event simulations (DES) use a mathematical model of a physical system, whose behavior can be described in terms of state changes [191]. A DES generates a "step-by-step succession of sets of events" in which each *set of events* denotes the (physical) system's state at a certain point in time [164, 30]. Hence, the physical system can be simulated with a sequence of discrete events (i.e., an event happens at an instant of (simulation) time). The developments pertaining to DES can be divided into four generations, which are described as follows [16]. First generation programming languages, like FORTRAN, have been used by programmers to manually define the simulation model logic and the control of the events. Then, special simulation languages appeared in the 1970s, like IBM®'s general purpose simulation system (GPSS), to increase the productivity for simulation model creation. In the early 1980s, simulation language generators further reduced the model development times. Around the same time, hardware description languages (HDL), encompassing a notion for events, were introduced to simulate electronic circuits (c.f. [259]). Over time, DES has been applied to a multitude of problem areas including weather forecasting, manufacturing, and health care.

Events have also been employed in distributed systems [69]. For instance, Lamport has described a distributed system as composed of several processes, where each process constitutes a *sequence of events* [154]. Initially, the inter-process communication was established by writing and reading values (messages) to and from a single (and later also distributed (cf. [77])) shared memory (cf. [273]). Agha described the ACTORS system in which concurrent processes in a distributed system exchanged messages in an *asynchronous* fashion [6]. Further, remote procedure calls (RPC), coined by Birrell and Nelson, have been introduced for inter-process communication [24].

Event processing also played an essential role in *networking technology* [164]. Networking started in 1969 with the ARPANET and led to the pervasiveness of the Internet. In any case, messages or network packets can be interpreted as simple events [164]. In that sense, network protocols, like TCP/IP, are employed to specify how packets (events) have to be exchanged (e.g., reliability, sequence) by senders and receivers. Further, the Open Systems Interconnection (OSI) model introduced a system of network layers constituting a hierarchical abstraction. In some sense, the OSI model has similarities to the event abstraction levels elaborated in Sect. 2.2.3.

Similar to the concept of *loose coupling* of software components[8], the module interconnection languages (MIL), which are better known as programming-in-the-large versus programming-in-the-small (cf. [52]), tried to describe the connection of software modules without considering their internal implementation [69]. Later, architecture description languages (ADL) have been used to model an information system's structure and interplay of components. For instance, the de facto standard unified modeling language (UML) encompasses the notion of *time events*, *call events*, *signal events*, and *change events* [195]. Also, RAPIDE, an event processing

---

[8] Loose coupling of (software) components is a main motivation for the use of EDA (cf. Sect. 2.2.5).

language invented at Stanford University, can be used to define models of system architecture [211]. Noteworthy, RAPIDE explicitly encompasses the notion of sets of *causally and temporally related events* [211]. Actually, Luckham, who has coined the term CEP, has extensively worked on RAPIDE, thus transferred insights gathered while working on RAPIDE to define CEP [161].

Moreover, events have been used in *graphical user interfaces* (GUI), starting with Apple® Macintosh™, Microsoft™ Windows®, and the X Window System developed at the Massachusetts Institute of Technology (MIT). The X kernel of the X Window System, for instance, has introduced the capability to capture events from the keyboard and mouse, and react to these events [242]. Further, Microsoft has provided Visual Basic, which is the first programming language with *built-in* support for events raised while using GUI elements (e.g., mouse click on a button) [69].

The development of *active databases* can be seen as more closely related to event (stream) processing. Traditional relational databases comprised an inability to monitor and trigger alerts and notifications [39], i.e., they exhibited a purely passive behavior. Consequently, events and rule concepts have been introduced in various databases (e.g., PostgreSQL) including traditional and object-oriented databases [39]. The use of event condition action (ECA) rules, especially became a salient feature. In recent years, *in-memory databases* have become pervasive as main memory has become less and less expensive. As event processing also exploits low latency of main memory, questions concerning the differentiation between in-memory databases and CEP are often raised (cf. [64, 231]). In the same way, in-memory databases are discussed as enablers for an RTE [18]. Hence, some concepts revolving around in-memory databases are discussed in the following paragraphs.

Accessing main memory (i.e., read and write operation) is orders of magnitude faster than accessing hard disk storage. Hence, as early as the 1980s, the idea to keep an application's entire amount of data in main memory emerged, primarily to achieve better response times and transaction throughputs [86]. The hard disks are then just necessary for backup and recovery purposes. However, an in-memory database organizes data according to the characteristics of main memory and the requirements of the considered application. Recently, a prominent in-memory database SanssouciDB, which has provided many concepts used in the High-Performance Analytical Appliance (HANA) of SAP [206] has been developed. A basic principle of this in-memory database is its column-based organization of data in main memory, which has advantages while performing operations on columns, like aggregations (e.g., derivation of an average value). Row-oriented and column-oriented organizations of data in (main) memory are contrasted in Fig. 2.9. In the case of the column-oriented organization, values of a column are stored in adjacent blocks in (main) memory, thus accessing values required for calculations, like sums, minimums, or maximums, are in orders of magnitude faster. Further, the memory footprint can be reduced employing data compression techniques [206]. Contrarily, reading a certain row representing an enterprise's entity (e.g., customer) is much slower when employing a column-based organization of memory. Summarizing, the row-based organization is mainly designated to be used for operational

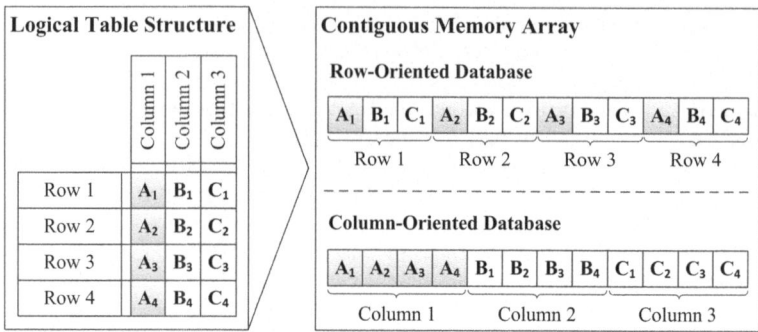

Fig. 2.9 Row- and column-oriented organization of a data table in a contiguous memory array (adapted from [206]).

data management, whereas the column-based organization has its strengths in analytical processing.

However, in any case, the interaction between a (in-memory) database with an (enterprise) system or a user, is primarily founded on a *request-reply pattern*, which is in clear contrast to the event-driven approach of EDA and CEP. Nevertheless, both concepts intend to offer real-time provisioning of relevant data, thus might be combined in some application scenarios.

Pursuing this discussion on distributed (computing) systems, various distributed component models [69] and middleware technologies became popular in the early 1990s. The main intention of middleware is to foster the development of distributed applications by abstracting network communication details. The Common Object Request Broker Architecture (CORBA), which has been standardized by the Object Management Group (OMG), was one of the precursors and also contains an event service, which has been further developed as a notification service [194]. The event service (later notification service) offers an asynchronous and loosely-coupled communication pattern for distributed software components [17]. However, CORBA started as a leading-edge technology, where main software vendors participated in its standardization, turned into being a niche technology [113]. Henning discusses CORBA's decline and advocates the theses that the technological flaws of CORBA have been caused by an insufficient standardization process [113].

In the mid-1990s, Microsoft^{TM} developed the Distributed Component Object Model (DCOM), which later got absorbed by the more successful .NET framework from the same vendor. Before the .NET framework, the Java Platform, Enterprise Edition (JEE, formerly known as J2EE) received a lot of attraction by developers, mainly because of its adherence to open standards and platform independence [121]. Both platforms have an accumulated market share of about 80 % [59]. Meanwhile, the SOA paradigm gained attention as a mean to achieve enterprise interoperability. Hence, both .NET framework and JEE support the development of SOAs based on web services.

Further, message queuing systems, also known as message-oriented middleware[9] (MOM), provide a *persistent asynchronous communication*, where neither a sender nor receiver must be active during message transmission [242]. In essence, a send message (conveying an event) is kept in a message queue until a predefined deadline is reached or a subscriber receives the message. Consequently, MOM systems are allowed to transfer messages in minutes instead of seconds or milliseconds [242]. Examples for MOM systems are TIBCO® Rendezvous, IBM® Websphere MQ or Microsoft™ Message Queuing [69].

Recently, events are regarded as central drivers within enterprise strategies, business applications, BPM, and the like. The term RTE, i.e., its drivers, characterization, and system requirements, has already been elaborated in previous paragraphs. In the same direction, an urgent need to handle events more effectively during execution of value creation process instances has been discovered. Consequently, the liaison between BPM and (complex) event processing is envisioned and has been labeled event-driven business process management (EDBPM) [14]. In subsequent paragraphs, EDA and CEP are presented as the state-of-the-art concepts or technologies pertaining to event processing.

2.2.5 Event-Driven Architecture

An EDA is one that enables IT systems to detect events and to react on them in an *intelligent* manner [243]. Therefore, events are considered as first class citizens of an EDA. An EDA is mostly described as composed of event producers, event processors, and event consumers. The basic structure of an event-based system that follows the principles of an EDA is depicted in Fig. 2.10. Events (i.e., something that happens or has happened) are produced within event producers. In this context, Taylor et al. have also introduced event listeners, which are capable of observing and interpreting the produced events [243]. For the sake of simplicity, event listeners have been excluded from Fig. 2.10. The observed events are transformed into a corresponding event object (i.e., notification). An event object is then published in a notification service. Event consumers that are interested in a certain (published) event, can subscribe to that event at the notification service. After receipt of an event within the notification service, the event is transferred by the same service as a message to the registered event consumers. The interested event consumers are notified by the notification service about the existence of a (new) event, which in turn can react on the receipt of event objects. Yet, it has to be clarified that the differentiation between event producers and event consumers is not always as strict as depicted in Fig. 2.10. Rather, it depends on the perspective if a (software) component is regarded as an event producer or event consumer. In fact, (software) components are oftentimes event consumers as well as event producers.

---

[9] The abbreviation MOM is also used for manufacturing operations management throughout this research work. However, the respective context wherein the abbreviation MOM is used makes clear which semantic is meant.

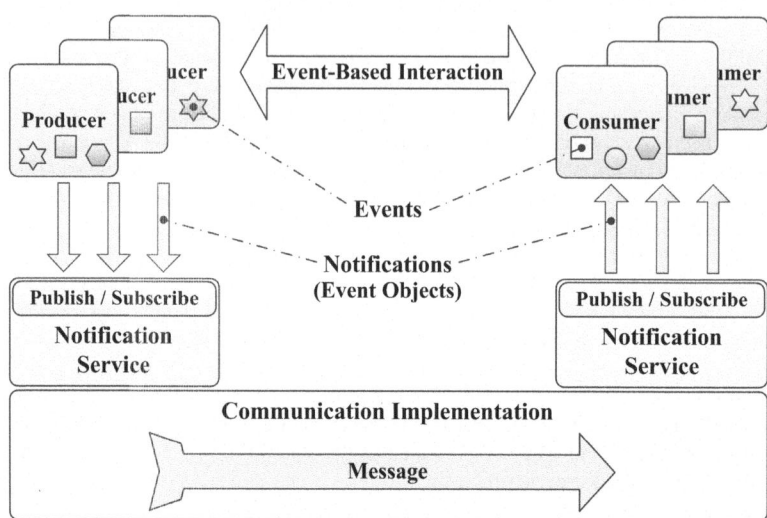

Fig. 2.10 Basic architectural design of an event-based system, which separates interaction from implementation (adapted from [190]).

Chandy and Schulte define a similar reference architecture for the EDA domain [41]. In their terminology, the communication implementation is provided as a *channel* (cf. also the communication implementations summarized in Fig. 2.11), and the notification service is just called as *intermediary*. Sometimes this intermediary is also called a service bus (e.g., enterprise service bus) [217, 243]. In any case, the intermediary (i.e., notification service) implements a *publish-subscribe interface*, comprised of advertisement, publish, (un-) subscribe, and notification operations [190]. The main effect of employing an intermediary is the *loosely coupled connection* between the event producers and event consumers. This connection is denominated as *event-based interaction* in Fig. 2.10. Noteworthy, the event-based interaction only exists at a *conceptual level* (i.e., between event producer and event consumer), and not at a *logical level* between event producer and intermediary or between intermediary and event consumer (cf. [41]). The transfer of notifications or event objects can be implemented by various protocols, as illustrated in Fig. 2.11. Some of these protocols have been described in the historical outline as predecessors and foundations of (modern) event processing (cf. Fig. 2.8). These protocols can be classified into shared resources and procedure calls, where the former are intended to transfer data and the latter for transfer of execution control [69]. The most widespread shared resources for conveying data are serialized connections, as in networking, where data is serialized into a sequence of bytes [69]. Procedure calls can be used for data transfer if they convey parameters. In the context of event-based systems, however, the focus is laid on shared resources, especially network connections.

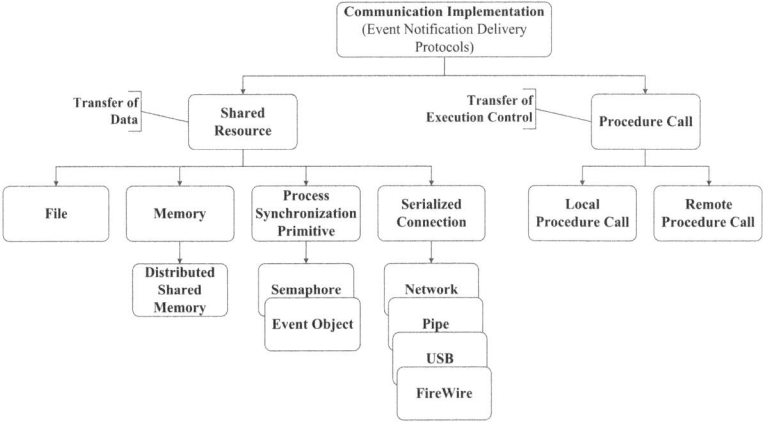

Fig. 2.11 Event notification delivery protocols for implementation of communication (adapted from [69]).

The event-based interaction style has to be differentiated from other communication patterns, which are also used in (distributed) systems. A taxonomy of co-operation models is given by Mühl et al. in Fig. 2.12 [190]. This differentiation is established by considering the initiator of a communication, and the addressing type used in the interaction. As a result, four basic interaction models can be described: (i) request-reply; (ii) anonymous request-reply; (iii) callback; and (iv) event-based.

The main property, which expresses the difference among the above interaction patterns, is the coupling between the involved components (i.e., event producer and event consumer). The coupling of components can be circumscribed with the term preconception, which is the "amount of knowledge, prejudice, or fixed idea that a piece of software has about another piece of software" [243, 68]. Further, coupling of (software) components is negatively correlated with the flexibility and maintainability of a distributed component-based system. Coupling can be static (i.e., hard-coded) or it exists dynamically (i.e., component dependencies are established at runtime of a system) [69]. In general, coupling increases the complexity of the overall system. Nevertheless, it has to be stressed that no system exists that completely eludes coupling of components [69]. Following, the above interaction models are described in light of coupling between the event consumers and event producers.

The request-reply pattern is often used in client server architectures, where a client initiates the communication by directly requesting data or functionality on the server side. The server, in turn, replies with sending data or performing a task. The direct addressing results in a *tight coupling* between the server and the client [69]. The request-reply principle is the prevalent interaction mode in (traditional) SOAs [217]. The services within a SOA are (relatively) independent and implemented based on standards, which leads to a loose coupling of components (i.e., services) at a technical level. However, at a conceptual level, the services are not decoupled as the initiating system (still) requires knowledge about which services to call [217].

Contrarily, instead of directly addressing the server, a client can also send a request to an arbitrary, dynamically determined set of servers [190]. Similarly, for instance, a client can first contact a directory service to resolve the address of a server, which can provide the requested functionality or data [69]. Directory services are widely used in the realm of SOA, where Universal Description, Discovery and Integration (UDDI) services are employed to dynamically discover other (web) services. Although this pattern adds more flexibility to the components' interaction, the dependency among the involved parties remains at a conceptual level [190]. Further, an *inversion of control* can be achieved, so that (event) producers become

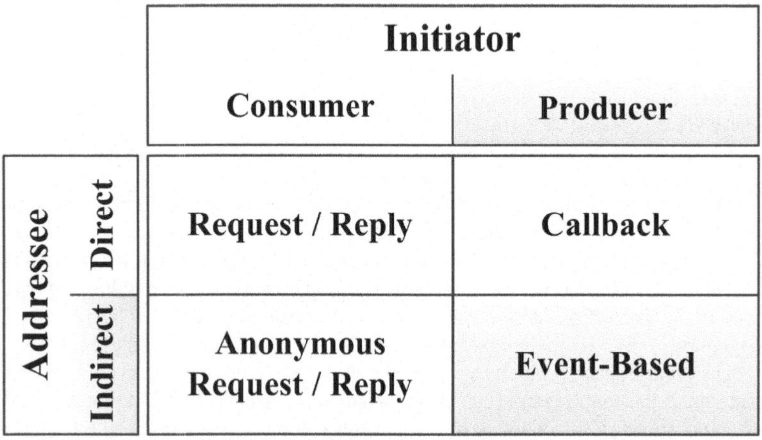

Initiator		
	Consumer	**Producer**
Addressee **Direct**	**Request / Reply**	**Callback**
Indirect	**Anonymous Request / Reply**	**Event-Based**

Fig. 2.12 A taxonomy of interaction models (adapted from [190]).

initiators of a communication, and thus, (event) consumers are notified by (event) producers. In that interaction pattern, a consumer *subscribes* to a producer that *publishes* some information or events, which are of interest to the consumer. The consumer supplies a reference to a *callback function* during subscription. The callback function can be called by the producer. If the information provided by the producer changes or an event is produced within the producer, then the consumer receives a notification. The notification either incorporates sufficient information (about the event) for the consumer (i.e., push model), or the consumer can request the (event) producer for further information (immediately) after receipt of the notification (i.e., pull model) [85]. A synonym for the publish-subscribe pattern is the observer pattern, in which an observer (object) has to synchronize its state with the state of a subject (object) [85]. Gamma et al. argue that the publish-subscribe (or observer) pattern averts a tight coupling between objects (i.e., either producer and consumer or observer and subject) [85]. In contrast, Mühl et al. stress that the identities of the communication partners have to be known and managed on both sides, and therefore, still result in a tight coupling of producer and consumer [190].

The evolution of the observer or publish-subscribe interaction pattern is an event-based interaction model. As elaborated in Fig. 2.10, the event-based interaction is accomplished by the insertion of an intermediary (e.g., service bus). Thus, the (event) producers and (event) consumers are *extremely loosely coupled* as they do not have any preconception about each other (cf. [243]).

An alternative classification of interaction models is given by Chandy and Schulte [41]. In the *time-driven interaction*, a communication or action is initiated at certain (physical) time instances by some software component. An example is the execution of a nightly batch job within an IT system. Next, the *request-driven interaction* is performed in client-server systems, as elaborated before. Finally, the *event-driven interaction* is initiated by an event, and it establishes a communication among the partners in an open-ended set of communication partners [41]. Noteworthy, wherever *celerity* is required in communication and processing (of events), time-driven and event-driven interactions are favorite models [41]. Nevertheless, every interaction style has its right to exist, as it is simply a matter of the business requirements. Also, for instance, SOA and EDA are complementary concepts for achievement of flexibility and maintainability. Summarizing, an EDA, which employs event-based or event-driven interaction styles, has the following properties, amongst others:

- As already stressed, the software components in an EDA are *loosely coupled.* Sometimes, the coupling is even described as being minimal or extreme (cf. [41, 217]).
- The communication between the event producer and event consumer is *unidirectional* [217]. This approach is also known as *fire-and-forget* [69].
- Because of the communication via an intermediary, an EDA does not rely on a *central controller* [243], that is, an (event) producer does not define or care about the (re-) action that should be or is taken by an (event) consumer [41].

An EDA can be used to separate a (IT) system in specialized distributed subsystems. An EDA can especially realize a sense-analyze-respond approach [219]. First, sensor systems observe and detect (internal and external) events as they happen. Second, these events are analyzed intelligently. Finally, on detection of a critical situation, appropriate (re-) actions are deduced.

2.2.6 Complex Event Processing

IT systems for intelligent analysis of events can be grouped into simple event processing (SEP), ESP, and CEP [184]. In a SEP system, the observation of a critical event directly results in a corresponding (re-) action. In an ESP system, events of a certain event type are partially ordered (e.g., partial order of events by their arrival times), which is capable of detecting patterns in the partial order of events. The most sophisticated form of event processing is delivered by CEP, which deals with the detection of complex event patterns in the event cloud introduced in Sect. 2.2.3. Also, Taylor et al. advocate CEP as an indispensable part of a modern EDA [243]. Thus, in the next section fundamentals and concepts of CEP are elaborated.

2.2.6.1 Basic Structure of a CEP Solution

Event producers, event processing, event consumers, and event distribution mechanisms constitute a CEP solution, as illustrated in Fig. 2.13. The event producers, event processing, and event consumers are interpreted as parties within an EDA, which are minimally coupled. Events are generated by machines, workflow systems, database systems, humans through human interaction systems (e.g., messengers, terminals), and the like [67]. These observable events are forwarded in an event-based fashion to complex event processing components that, in turn, *analyze* the events and their relationships. This event analysis is conducted by event processing agents (EPAs), which are interconnected in event processing networks (EPNs). Some EPAs are linked to input adaptors that are interfaced to the event distribution layer. On the opposite side, EPAs can be connected to output adapters, which forward appropriate deduction (information) to event consumers. The above schema of a CEP solution represents a sense-analyze-respond approach [32]. In this context, Chandy and Schulte differ among manual, partially automated, and fully automated CEP [41]. Manual CEP is performed by humans that take (re-) action after analyzing various events. In the case of partially automated CEP, a CEP engine is used to assist humans to make proper decisions. Finally, a fully automated CEP solution senses events, processes them, and automatically dispatches commands to influence business processes, software, hardware, and the like. In the following section, the constituents of the event processing part of CEP are elaborated.

2.2.6.2 Event Processing Agents and Event Processing Networks

As noted in Fig. 2.13, EPAs are central components within CEP. The term EPA was first coined by Luckham as "an object that monitors an event execution to detect certain patterns of events" [161, 176]. The basic structure of an EPA consists of an input terminal, event processing rules, and an output terminal [161]. The rules of an EPA are used to describe its behavior [67, 161].

The functioning of an EPA can be separated into a sequence of three steps [67]: (i) a filtering step, which selects events that (should) take part in event processing; (ii) a matching step, which detects patterns of events (i.e., set of events); and (iii) a derivation step, which performs (mathematical) operations on the detected event set, and further, generates output events. Here, it is not necessary to execute all steps of this sequence, that is, for instance, some EPAs are just used for filtering. EPAs are networked in EPNs that, in turn, are organized in a hierarchy of EPNs [161]. The main aim of the interconnected EPAs is the detection of event patterns in the event

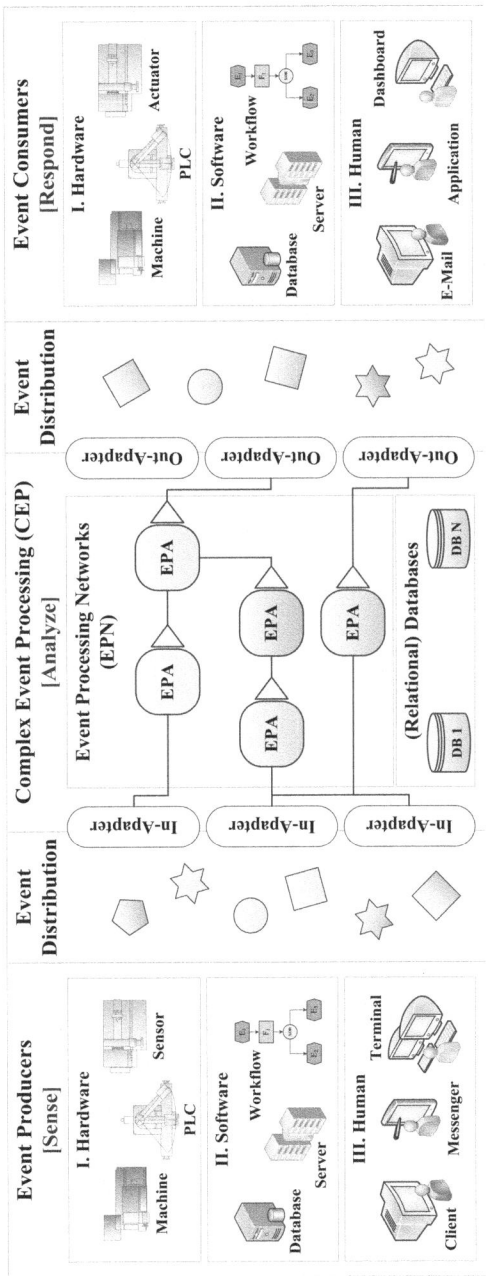

Fig. 2.13 Basic architecture of a CEP solution, which follows a sense-analyze-respond approach.

cloud, i.e., they are a means to realize the event abstraction hierarchy delineated in
Fig. 2.6.

A classification of EPA types is provided by Etzion and Niblett in Fig. 2.14 [67].
A *filter* EPA is a simple EPA type that just selects events for subsequent processing steps. *Transformation* is an abstract EPA type that can be represented as a (i)
translation; (ii) aggregation; (iii) splitting; or (iv) composition EPA. *Enrichment* and
projection are translation EPAs. The former is employed to add or update event attributes to an event, e.g., by assigning values taken from a (relational) database (cf.
Fig. 2.13). This database (or other persistent memory) represents a global state
that encompasses historical events, reference data, and the like [41]. A *projection*

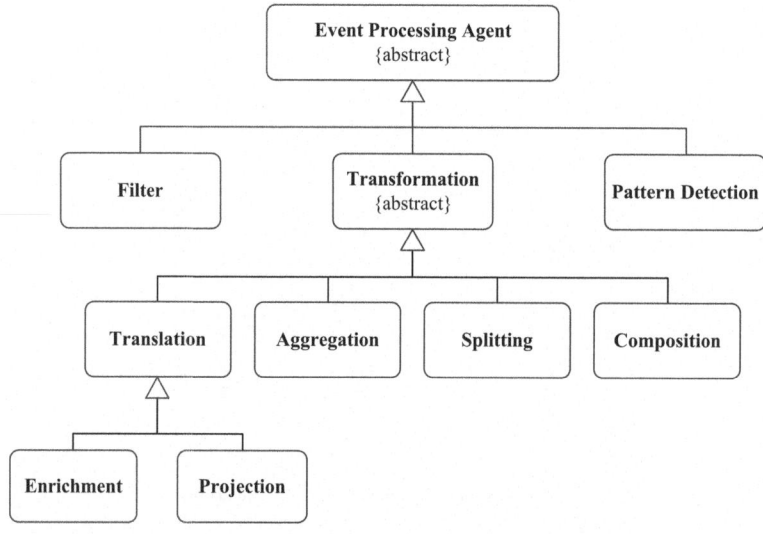

Fig. 2.14 Classification of event processing agent types (adapted from [67]).

EPA takes a subset of event attributes of a single event and uses them to generate
a derived event. Event *aggregation* is also known as *event pattern mapping*, where
events are combined or correlated to create complex events [161]. Special derivation functions, like sum, average, or maximum, are used to derive aggregated event
attributes based on a set of event attributes taken from the input events. A *splitting*
EPA divides a single event into two or more events. A *composition* EPA can join
separate events based on a matching criterion (e.g., match of primary and foreign
keys). Finally, *event pattern detection* is performed on event streams to detect (complex) patterns by analyzing causal, temporal, and other event relationships (cf. Sect.
2.2.3). Luckham paraphrases event pattern detection as event pattern constraint that
is a special type of event pattern map [161].

In general, an EPA can be *stateful* or *stateless* based on the processing of a single event or a set of events, respectively [67]. Further, EPAs can take an *event context* into account. An event context is a means to assign an event into one or more context partitions (i.e., event groups), which can be temporal, segmentation-oriented or state-oriented [67]. For instance, in the case of a segmentation-oriented context, for each individual context segment an instance of an EPA is built for processing of events that are members of this context segment. Temporal context is often expressed using a (sliding) window concept, where a window contains only events whose (arrival) timestamps are between a window start and window end (e.g., timestamps) [39].

In addition to the detection of event patterns, an EPA can deduce (re-) actions. In its simplest form, an EPA just emits a derived event that can be consumed by other EPAs or an event consumer. Further, an EPA can call a service of an application system (i.e., a service within a SOA, method or function call) that, in turn, influences a (business) process, updates the visualization of a chart within a dashboard, and so forth [32].

2.2.6.3 Event Processing Languages and Event Processing

As mentioned earlier, an EPA's core contains event processing rules that are executed by a CEP engine. These rules are written using an event processing language (EPL) [67], sometimes also called event query language (EQL)[10] [61]. The main advantage of employing EPLs is the separation of the event processing logic from the application logic [67]. Consequently, it is not required to (re-) compile a CEP solution after modification of the event processing logic, thus leading to enhanced CEP system flexibility. From a technical perspective, main design goals for an EPL are a *high throughput* of events, *low* reaction or *response times* (i.e., latency), and at the same time, an execution of *complex event analysis operations* (e.g., event pattern matching, event correlation, event aggregation).

Thus far, there exists no standard EPL, either in science or in CEP markets [61, 67]. Hence, the EPTS has founded a working group to analyze various event processing language styles. A first outcome of their work was presented at the Distributed Event-Based Systems (DEBS) Conference in 2009 (cf. [34]). In summary, the following language styles can be identified (cf. [34, 61, 67]):

1. Rule-based languages that might be subdivided into

   a. production rules
   b. active rules
   c. logic rules

2. Imperative programming or scripting languages
3. Stream-oriented languages following the structured query language (SQL)
4. State-oriented languages

---

[10] In subsequent paragraphs, the terms EPL and EQL are used interchangeable.

In general, the above mentioned language styles were preferred in other application fields long before they got applied in CEP. Therefore, the basis of the EPL is not an original outcome of the CEP community. The existing language styles have been adapted or extended to fit the needs of CEP. The extensions are primarily introduced (i) to cope with temporal relationships between events; and (ii) to detect complex event patterns. Following, each language style is explained independently. However, in practice, CEP solutions make use of and combine several language flavors.

Production rules[11] cannot be designated as EQLs in a narrow sense [61] as they react on state changes and not on events directly [34]. The basic scheme of a production rule is: "IF *Condition* THEN *Action*." Conditions are defined over objects (e.g., plain old Java objects (POJO)), which are also known as facts. These facts are added to a fact base or working memory that, for instance, is implemented as a (event) queue. After an event (i.e., event object) has been added to the working memory, the rule engine has to be fired for evaluation of the rules (i.e., all rules are fired until no rule can fire) [61]. In most instances, the rule evaluation (i.e., forward chaining) is based on variants of the RETE algorithm that was developed by Charles Forgy and published in 1982 [79]. Production rules are pretty flexible, but can also entail a low abstraction level resulting in an impaired expressiveness and reduced ease of use [61].

ECA rules were first introduced in active databases (cf. Sect. 2.2.4). Summarizing, the building blocks of the ECA approach can be described as follows: "Events correspond to *happenings of interest*, conditions correspond to *ascertaining the validity of a state*, and actions correspond to the *outcome of the situation*" [emphasis in original] [39, 42]. Further, *logic programming rules* can be used for event processing, as they are employed in ETALIS [66,67]. Logic languages possess a strong formal foundation, are highly expressive, and their logic rules can be specified in a convenient manner [61].

Overall, the rule-based approach for EPLs is widely used in industry (e.g., TIBCO BusinessEvents [249]). The vendors of rule-based systems have addressed CEP requirements by extending their rule-based systems (RBS)–more precisely, their rule languages–to take temporal relationships, derivation of aggregations, support for multiple concurrent (event) streams, and the like, into account (cf. [188]).

In computer science, programming languages can be roughly differentiated as declarative and imperative [220]. The rule-oriented, stream-oriented, and state-oriented languages can be classified as *declarative languages*, that is, a query describes *what* should be the result of the event processing. In contrast, *imperative languages* precisely describe *how* to achieve a result, that is, for example, a sequence of operations or steps is defined. In the context of CEP, imperative EPLs focus on the response (i.e., action) part of CEP (cf. Fig. 2.13). An (sophisticated)

---

[11] The term *production rule* is used differently in the realm of manufacturing operations management (MOM). There, it is defined as "information used to instruct a manufacturing operation how to produce a product" [126, 12].

action can be formulated employing a script language as is the case, for instance, with Progress Apama[12] Monitorscript [160].

Nevertheless, the focus in this research work is based on *stream-oriented languages*. The stream-oriented languages are primarily inspired by relational algebra. Stream-oriented languages are optimized with regard to low latencies and are known by many users. However, the processing model of event (data) in a CEP or ESP solution differs substantially from the one that is used in regular (relational) database management systems (DBMS). Chakravarthy and Jiang describe the paradigm shift that has led to dedicated data stream management systems (DSMS), where data (i.e., events) have to be analyzed in a continuous and timely manner [39]. They also describe that DSMS can be integrated with CEP. In other words, DSMS–especially in the form of ESP–is often the primary building block for stream-based CEP solutions (cf. the discussion of ESP and CEP in Sect. 2.2.3).

Basically, DSMS exploits the relatively low latencies of main memory in contrast to high latencies of hard disks[13], i.e., the processing of events is completely performed *in-memory*. In short, a DSMS or CEP solution can be considered as a regular DBMS that has been *turned upside down* [39]. In a DBMS, the *data are persistent* and the queries are performed ad-hoc, i.e., *queries are volatile*. Contrarily, in a DSMS, *queries are persistent* and *transient data* is streamed through these queries continuously.

Overall, the in-memory processing of event streams incurs specific challenges (cf. [39, 150]):

- Some stateful operators, like join, sort, and aggregation, cannot produce a result as long as their corresponding stream has not been terminated. Accordingly, these operators are called *blocking operators*.
- There is no random access to the input events. Rather, the *access* has to be organized *sequentially*.
- In contrast to DBMS, main memory is a bounded resource, thus leading to a computation based on a *limited amount of space*.
- The processing has to be performed using fewer computing cycles, that is, results should be *derived continuously* (i.e., on-the-fly).

A comprehensive discussion on special characteristics of DSMS in contrast to traditional DBMS is given by Chakravarthy and Jiang (cf. [39]). For the sake of completeness of the previous discourse, the outcome of their discussion is illustrated in Fig. 2.15. In the next paragraphs, common concepts of DSMS that have been developed to address the aforementioned challenges of DSMS are elucidated, and thus, explain their characteristics mentioned in Fig. 2.15. To achieve a non-blocking behavior of (stateful) operators, the concept of (sliding) windows has been introduced for (continuous) queries used in DSMS. A window defines a finite portion of a stream of events [39]. There are several different window types (e.g., depending

---

[12] Progress Apama is a CEP solution, which is primarily focused on the financial industry. See also their web page: http://www.progress.com/en/apama/

[13] Recall, that in-memory databases have a similar motivation (cf. Sect. 2.2.4), but (still) follow a request-reply pattern (cf. Sect. 2.2.5).

Database Management System (DBMS)	Data Stream Management System (DSMS)
Persistent relations	*Transient* streams
One-time (*ad hoc*) queries	*Continuous* queries
Random (*hard disk*) access	Sequential (*in-memory*) access
Unbounded disk storage	*Bounded* main memory
Snapshot of current state used	Arrival order, *window* important
Relatively *low update frequency*	Varying and *bursty input rates*
No Quality of Service support	Quality of Service support critical
Requires *precise results*	May tolerate *approximate results*
Computes all results on a *snapshot*	Computes window-based results *incrementally*
Transaction management critical	Transaction management not critical

Fig. 2.15 Comparison of characteristics of database management systems (DBMS) and data stream management systems (DSMS) (adapted from [39]).

on the specification of the start and end boundaries of a window), of which the sliding windows are most rampant. The length of a sliding window can be defined by specifying the number of events or a particular timespan.

An example of a sliding *time* window is illustrated in Fig. 2.16. Incoming events are kept in main memory as long as their arrival (or occurrence) timestamp is within the interval of the current system time *minus* the time-span specified for the sliding time window. Events that exceed this condition are pushed out of the window (i.e., outgoing events). Thus, the number of events in the sliding time window is not limited to a fixed number (i.e., the width or range of the window), and usually varies over (physical) time. In contrast, a sliding *length* window defines a maximum number of events that should be kept in the sliding window; an example is illustrated in Fig. 2.17. In addition to these window types, there exist hybrid window types, i.e., a combination of time and length windows. In general, the concept of (sliding) windows is a fundamental concept in data or event stream processing. On one hand, windows limit the number of events that have to be processed in an operation (e.g., sum, average), therefore they enable non-blocking operations. On the other hand, they store records non-persistently in a short-term memory (i.e., in the main memory of the host system). A window can be considered as the counterpart of a relation in a traditional (relational) database.

More widely, the notion of *snapshot-reducibility* has been introduced for the extended relational algebra used in continuous queries of DSMS [150]. In short, a stream operation is snapshot-reducible if its relational counterpart produces the equal result [150]. Hence with the introduction of windows, stream-oriented languages offer the same expressiveness as SQL-like languages. In many instances, it is also possible to merge an in-memory data stream with a relation stored on hard disk (cf. enrichment EPA in Sect. 2.2.6.2).

Example: `SELECT * FROM SensorData.win:time(3 seconds)`

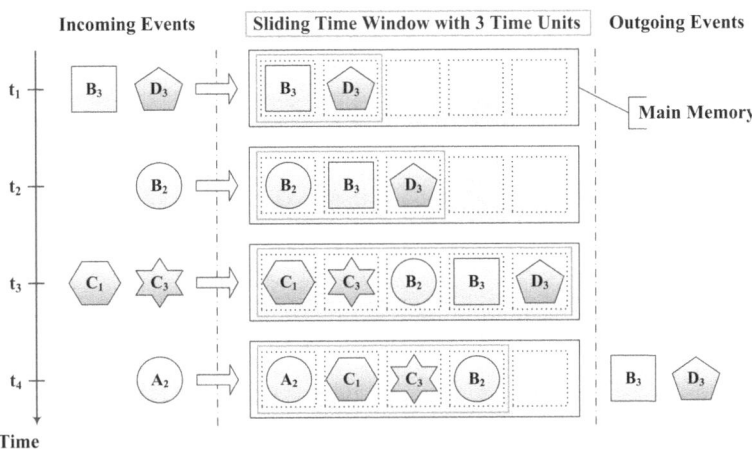

Fig. 2.16 Sliding *time* window with window timespan of three time units (adapted from [32]).

In technical terms, a formulated stream-oriented (or continuous) query can be conceptualized as a dataflow diagram [39]. Each node represents a (logical) operator (e.g., select, sum) and the edges are events flowing between the nodes. Especially in front of stateful operators, (main memory) queues are implemented to store unprocessed events [39]. The result of an operator is written to a queue in a non-blocking fashion, but read in a blocking fashion [34]. Further, a query plan in a DBMS is traversed top-down and left-to-right, thus is executed in a pull-based mode; a query plan in a DSMS is executed in a push-based mode starting from leafs and travelling up to the root.

So far, the extended relational algebra part of stream-oriented EPLs has been described. Nevertheless, the detection of *event patterns* is a feature of CEP solutions, which is of paramount importance. Therefore, stream-based EPLs have been extended with statements to express certain event patterns[14]. Specifically, the specification of sequences of (non-) events has been integrated into EPLs. The detection of complex/composite events can be implemented with event detection graphs (EDG) based on event algebra, colored Petri nets, or *extended finite state automaton*, also called extended finite state machines (FSM) [39]. An example of an EPL that specifies a temporal event pattern is illustrated in Fig. 2.18. This event pattern detects if an event of event type A or B is *followed by* an event of type C *within five minutes* after observation of the event of event type A or B. To support temporal event

---

[14] An overview of various event patterns is given by Etzion and Niblett [67]. The event patterns can be assigned to pattern groups (cf. [67]), like (i) basic patterns; (ii) threshold patterns (e.g., count pattern, max value pattern); (iii) temporal patterns (e.g., sequence, increasing); (iv) spatial patterns (e.g., minimum distance), and so forth.

Example: `SELECT * FROM SensorData.win:length(5)`

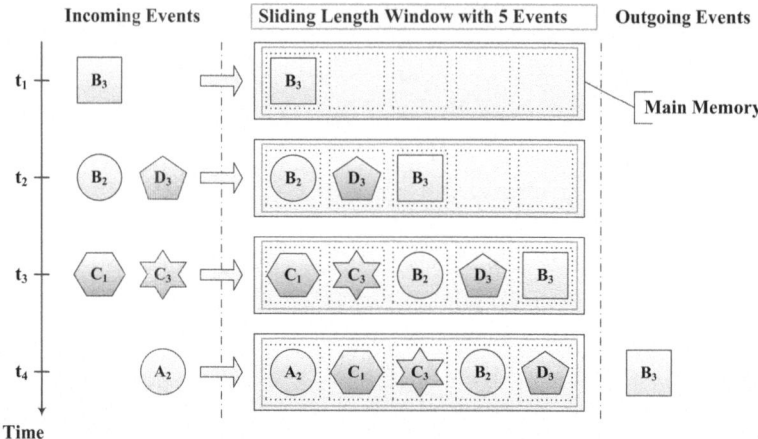

Fig. 2.17 Sliding *length* window with a maximum of five events (adapted from [32]).

sequences, an FSM has to be extended by a generative time state. Once a time state has been entered, it will be purged after a predefined time-span expires. As known from theoretical computer sciences, deterministic and non-deterministic finite automata (DFA and NFA) define a regular language, thus are used, e.g., in compiler construction for detection of tokens (cf. [7]). Actually, the event patterns expressed in EPLs have the expressiveness of a regular language (i.e., language of type three in the Chomsky-Schützenberger hierarchy). The main difference is the use of events instead of characters of an alphabet as state transitions.

Timed state automata (cf. Fig. 2.18) are regular FSMs, which have been extended to support temporal aspects [61]. They can be considered as a separate EPL style, but are usually integrated with other EPL styles, as illustrated in Fig. 2.18. In isolation, timed state automata cannot express event aggregations, and modeling of event negation and event composition is cumbersome [61].

2.2.6.4 Modeling of (Complex) Events

Although (commercial) CEP solutions provide some means for graphical modeling of event processing (EP) statements, the underlying EPLs are textual and not graphical. Contrarily, graphical modeling is common in the realm of business process reengineering (BPR) and BPM. EPC, BPMN, and the like, and are established modeling languages for business processes. However, the transitions between process steps (i.e., functions) of a business process are primarily modeled as simple events (cf. Sect. 2.2.1), thus complex events are not supported [216]. The trend to couple

I. Event Pattern Specification:

```
select C from pattern [ every (A or B) ->  C
where timer:within(5 min)]
```

II. Verbal Description of Event Pattern:

Detection if an event of event type A or B is *followed by* an
event of type C *within 5 minutes* after observation of the event
of event type A or B.

III. Finite State Machine (FSM):

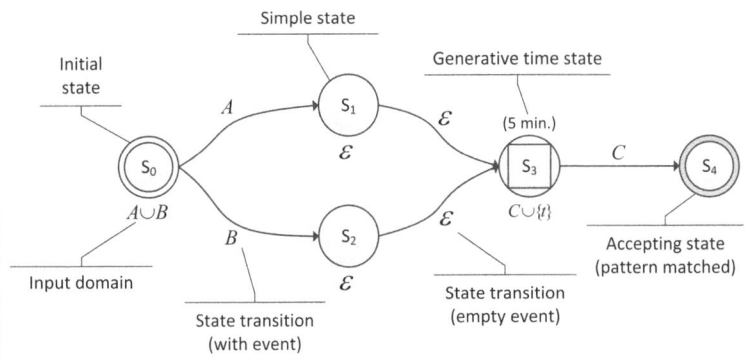

Fig. 2.18 Complex (or composite) event detection realized as a finite state machine (adapted from [190]).

business processes with the physical world using sensor networks and RFID[15] has disclosed limitations to model complex events in BPM [51].

To overcome this situation, Rommelspacher presents an approach to model aggregated events as *event diagrams*, where (simple) events are illustrated with the EPC event symbols and are interconnected with logical operators [216]. Although this approach is useful in some instances, it basically neglects temporal relationships between events. Hence, decision tables are also introduced to express more sophisticated event relationships. These decision tables are linked to a special event connection operator. In addition, a dedicated Business Event Modeling Notation (BEMN) has been presented (cf. [51]). This notation encompasses (i) a specification of graphical elements for modeling of complex events; and (ii) an abstract syntax of BEMN models to provide formal execution semantics [51]. Examples, following BEMN, are modeled in the presentation of industrial use cases in Sect. 5.4.4.

---

[15] Notably, the fusion of the virtual and the physical world has been discussed by Fleisch and Österle as a main characteristic of the RTE (cf. [78]).

Chapter 3
Requirements Analysis

Fundamentals and principles of the RTE and its application to manufacturing enterprises were *introduced* in Chap. 2. The main building blocks of an RTE (i.e., (vertical) integration, (process) automation, and individualization) have been discussed in general. The RTE has been motivated as an appropriate strategy to address business pressures (cf. Fig. 2.2). As a result, (manufacturing) enterprises strive for achieving celerity, agility, and information availability. As outlined in the introduction of this research work (cf. Fig. 1.1), the transmission of the RTE vision to manufacturing enterprises is affected by managerial, engineering, and information technological issues. Henceforth, problem perspectives, methodologies, and solution approaches of the management, engineering, and computer science community are elaborated in more depth. This presentation reveals the interdisciplinary character of the presented research work that is typical for many problems and methodologies presented by the ISR community. In addition, the application area for implementing/testing the presented research work is described. Thereby, from the perspective of philosophy of sciences, the slice of reality/universe of discourse is disclosed.

3.1 Description of Application Area

In short, this research work is focused on enterprises that (i) create value primarily by *manufacturing* tangible products; (ii) are of small or medium size (i.e., belong to the group of small and medium-sized enterprises (*SMEs*)); and (iii) contribute to the German national income as part of the *German industry*. Before delving into the aforementioned problem perspectives, some light is shed on the traits of these enterprises.

3.1.1 Manufacturing - Economic Relevance and Classification

Manufacturing can be defined as a combination of *technological processes* to alter the geometry, properties, and appearance of a starting material to make products [105]. Kumar and Suresh define production[1] as "the step-by-step conversion of one form of material into another form through chemical or mechanical process to create or enhance the utility of the product to the user" [15, 7]. In addition, manufacturing can be understood as an economical process, where value is added during transformation of input materials/products [105]. Usually, the assumed definition is clear in a given context. Hence, the economic relevance of manufacturing in the German economy will be discussed.

The production of services and goods has been organized in economics traditionally into primary production (i.e., mining and agriculture), secondary production (i.e., manufacturing), and tertiary production (i.e., services). Sometimes, a quaternary sector (i.e., information or intellectual activities, like scientific research and education) as well as a quinary sector (i.e., highest level of decision making in a society and economy) are mentioned also (cf. [218]). Measured in terms of value added to German national income, these sectors possess different shares in the economy.

Research has been carried out on structural changes, i.e., changes pertaining to the composition of an economy (cf. [151] for a concise survey on structural change)[2]. Actually, in developed countries, like Germany, the tertiary sector has gained importance relative to the primary and secondary sectors [151]. In the German economy, the tertiary sector encompasses information and communication technology (ICT), logistics, trading, financial industries, and the like, and about 69 % accounts for the biggest share on the German national income [74]. Contrarily, the secondary sector, or manufacturing sector, is responsible for about 25.6 % of the German national income [74]. The above structural change in the German economy can also be illustrated with the share of total employment in various economic sectors, as depicted in Fig. 3.1. Despite the structural change and the strong growth of the tertiary production (i.e., services), secondary production (i.e., industry, manufacturing) remains important for the German economy. Similarly, for instance, the BMWi argues that German industry is still the backbone of the German economy, and is mainly responsible for the demand for services offered in the tertiary sector [72]. Similarly, a study on the relationship between the service sector and the industrial sector reveals that around 40 % of production activities (i.e., both in the secondary and tertiary sector) in the German economy are indirectly induced by demand for material products (i.e., tangible goods) [134]. Consequently, Kalmbach et al. put the thesis of German de-industrialization into question [134].

---

[1] Production can be interpreted as a wider term than manufacturing (e.g., production in the service industry) [104]. Nevertheless, the terms manufacturing and production are used interchangeably in the context of this research work. Except that the context implies a wider meaning of the term production, it is understood as industrial production (i.e., manufacturing in the narrow sense).

[2] For instance, the three-sector hypothesis formulated by Colin Clark and Jean Fourastié is one of the most prominent descriptions and explanations regarding structural change (cf. [45, 81]).

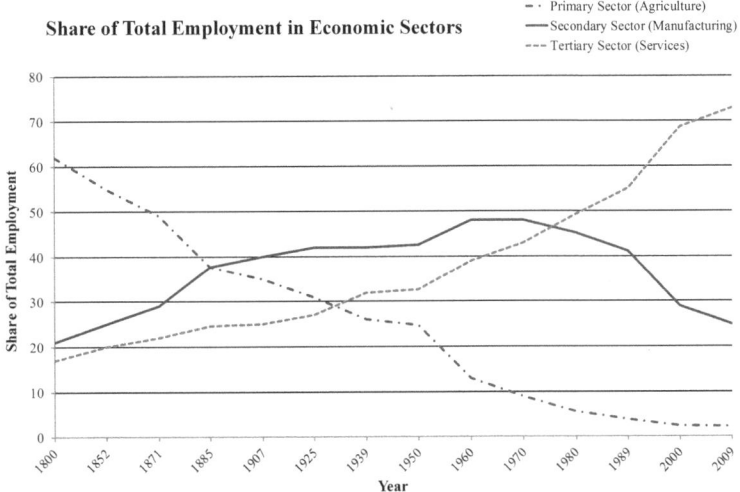

Fig. 3.1 Structural change in German economy illustrated with the share of total employment in economic sectors (adapted from [88]).

In contrast to other industrial nations[3], the German economy largely relies on industrial production [3]. German industries, like automotive, machine building, and so forth, are responsible for about 76 % of expenses in innovation projects [209]. Noteworthy, automotive, machine building, chemical industry, metal processing, and electronics have a share of around 57 % of the German exports [73]. And about 90 % of the German exports are tangible goods [3]. However, the strict separation of tangible products and intangible services has been blurred as both are combined in hybrid value creation processes[4]. These processes have gained more importance in industry as significant selling points, and thus, are extensively researched (cf. [28, 149]).

The production of tangible goods can be differentiated using various aspects that are product-oriented, production-oriented, marketing-oriented, and the like (cf. [48, 70, 106]). A (relevant) selection of properties to specify a manufacturing enterprise, manufacturing systems and manufacturing processes is presented in Fig. 3.2. Each property on the left side is numbered and can have up to three characteristic values. In addition, the scopes (i.e., high, medium, and low) of each property according to

---

[3] Abele and Reinhart compare Germany with Japan, Italy, the United States of America (USA), France, and Great Britain (cf. [3]).

[4] An overview of terms, which have been used in literature, with regard to the combination of products and services, like hybrid product, integrated solution or product service systems, is presented in [251].

the research work are highlighted in Fig. 3.2. The properties, which are assumed in this research work, are listed as follows:

1. The labor, (physical) resources and materials have a medium or *high degree of specialization*. For instance, workers have a high standard of education and are trained to do special operations. The employed machines are specialized for some operations, thus are designed to achieve higher productivity. Also, input materials are adapted for special products.
2. The considered manufacturing systems and manufacturing processes are *highly automated*, i.e., they have a low manning level [105]. The processing operations, assembly operations, and logistic operations are performed primarily by machines and automation devices.
3. The products are manufactured in series (lot sizes) that are of *low volume*.
4. The *variety of product types* is relatively high. Hence, there is a clear demand for advanced control of manufacturing in terms of performance, quality, inventory, and maintenance. The increased variety of products is one of the results of the business pressures mentioned in Fig. 2.2, and it is a motivation to address individualization (i.e., mass customization) in the RTE (cf. Sect. 2.1 and [78]).
5. The manufacturing of products is triggered by customer orders, i.e., the operational objective can be denoted as *make-to-order (MTO)*. This can also include the engineering of products before the start of manufacturing (i.e., *engineer-to-order*).
6. The plant layouts, which are focused on in this research work, mainly follow product layout (e.g., flow line) or cellular layout (i.e., *cellular manufacturing*). In a *flow line* production, multiple machines are arranged in a sequence, and (semi-finished) products are moved through these machines [105]. *Mixed-mode* production lines, which are capable of manufacturing various different product types, are assumed in this research work [105]. In contrast, in *functional manufacturing*, machines are grouped in departments according to their functionalities (e.g., lathes, milling). Such a plant layout is also described as a process layout. In scheduling research, flow line production is also called *flow shop production*, while functional manufacturing is denominated as *job shop production* (cf. [204]). In addition, in *cellular manufacturing*, a (production) cell is composed of several machines that are employed to execute different process operations to produce a (semi-finished) product [105]. The cellular layout tries to retain the flexibility of job shop production and concurrently exploits the efficiency of flow shop production [245]. Typically, nearly all the above mentioned layouts can be found in an industrial enterprise's plant [106].
7. The routing between different machines along the manufacturing processes is primarily fixed or predefined. Hence, questions concerning *logistics of material and (semi-finished) products* are not examined in depth. Similarly, Groover mentions that job shop production emphasizes production planning (scheduling and routing) whereas flow shop production focuses on production control [104].
8. The flow of material and (semi-finished) products can be both continuous and discontinuous.

9. The focus is on discrete manufacturing, i.e., the final products are discrete and uniquely identifiable, either physically or virtually identifiable.
10. *Intermediate* goods can be both bulk goods (e.g., liquids) and piece goods. For the sake of a proper traceability of the manufacturing process, both bulk and piece goods are identifiable.

1	Degree of Specialization of Elementary Factors	High Specialization	Medium Specialization	Low Specialization
2	Level of Automation	Fully Automated	Hybrid (Semi-Automated)	Manual (with feedback)
3	Production Quantity	Mass Production	Series / Serial Production	Individual Production
4	Product Variety	Low	Medium	High
5	Operational Objective	Make to Stock		Make to Order
6	Production Facility Type (Plant Layout)	Flow Line, Quantity (Product Layout)	Cellular Manufacturing (Cellular Layout)	Functional Manufacturing (Process Layout)
	Scheduling	Flow Shop		Job Shop
7	Routing	Fixed Routing		Variable Routing
8	Flow of Material	Continuous		Discontinuous
9	Industry Profile (Type of Manufacturing)	Process Manufacturing		Discrete Manufacturing
10	Shape / Structure of (Intermediate) Goods	Bulk Goods (e.g., Liquids, Batches)		Piece Goods
	Scope	High	Medium	Low

Fig. 3.2 Specification of manufacturing enterprises and their manufacturing systems as well as manufacturing processes, and definition of the scope of this research work (classification scheme mainly inspired by [105]).

3.1.2 Small and Medium Sized Enterprises - Structure, Traits, and Challenges

In addition to the aforementioned specification of production systems, information on enterprise size and ensuing implications with regard to the organizational and operational structure of a manufacturing enterprise are elucidated, especially, SME^5 are addressed in this research work.

According to the definition of the EU, an SME is an enterprise with fewer than 250 employees, an annual turnover of fewer than 50 million Euros and/or a balance sheet total of fewer than 43 million Euros [68]. A similar definition is provided by the Institut für Mittelstandsforschung (IfM), Bonn, Germany, where an SME has fewer than 500 employees and fewer than 50 million in Euro revenue [123]. Overall, 99.7 % of the enterprises in Germany are SMEs [125]. Further, family-owned enterprises, which can even have more than 500 employees, make up to 95.3 % of all German enterprises [124].

In the context of the presented research the following characteristics of German (manufacturing) SMEs are noteworthy (cf. [252, 254]):

1. The German SMEs, especially the ones that are world market leaders, follow a niche strategy (cf. also [264]) and are focused on operational effectiveness (e.g., effectiveness of manufacturing processes). German manufacturing enterprises focus on *monitoring management*, whereas U.S. manufacturing enterprises are excellent in incentive management (i.e., hiring, bonus, exit) [27].
2. In contrast to large enterprises, the organizational structure of an SME is relatively *flat* [253]. As a consequence, even the top management of an SME participates in innovation processes (cf. [155]).
3. More than 90 % of German world market leaders are SMEs [155]. The production of goods has been recognized as key for achievement of sustainable business success. In contrast to U.S. firms, market trends, like outsourcing and off-shoring (i.e., relocation of production to low-wage countries), are less popular for German (manufacturing) enterprises [155]. Rather, innovation processes that integrate research and development with domestic *production* have been established [155].
4. Driven by an enterprise's environment that can be circumscribed as being volatile and compelled to handle uncertainties (cf. also Sect. 2.1), manufacturing enterprises, especially suppliers, have to cope with a *high mix* and *low volume product mix*.
5. The majority of German manufacturing enterprises are SMEs, which are family-owned, i.e., ownership and leadership are united (cf. [110]). This trait has the following implications: (i) strategic and *operational independence* are of high value; (ii) self-financing is preferred; (iii) *job tenures* are usually long; and (iv) strategies are intended to be *continual* [253].

Keeping these traits of the considered manufacturing enterprises in mind, the management, engineering, and computer science perspectives of a manufacturing enterprise are elaborated. These perspectives are discussed in the context of manufacturing (e.g., its management, solution approaches, and the like). Further, the subsequent outline expands on the problem description elucidated in Sect. 2.1.

---

[5] Henceforth, the terms manufacturing enterprise, enterprise, and SME are used interchangeable throughout this research work.

3.2 Management Perspective

From a management perspective, *production management* "deals with decision making related to production processes so that the resulting goods or services are produced according to specifications, in the amount and by the schedule demanded and out of minimum cost" [35, 23]. The objective is to produce goods of the right *quality* and *quantity* at the right *time* with the right manufacturing *cost* (cf. [15]). The decisions in production management are influenced by decision relevance, amount of aggregated information available for decision-making, decision level and planning horizon, as illustrated in Fig. 3.3. Further, the decisions made in pro-

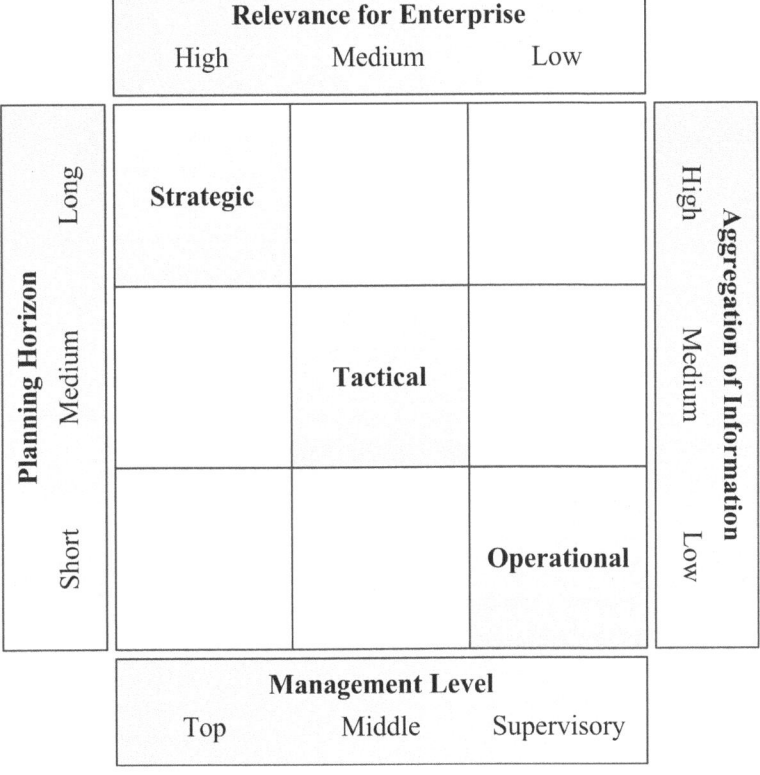

Fig. 3.3 Traits of strategic, tactical, and operational decisions in production management (adapted from [106]).

duction management can be classified based on (i) how relevant they are for the enterprise; (ii) how much aggregated information is employed in the decision making process; (iii) the management level at which the decision is made; and (iv) the planning horizon [106]. As a result, strategic, tactical, and operational decisions can be distinguished, as illustrated in Fig. 3.3.

Strategic, tactical, and operational decisions can be assigned to enterprise levels. These enterprise levels are depicted in Fig. 3.4 along with a few selected characteristics. The *top management* of a manufacturing enterprise defines the *enterprise strategy* for the achievement of enterprise objectives and to sustain its competitive advantages. The enterprise strategy defines guidelines, principles, and policies of the enterprise. Also, *enterprise visions* are defined at the strategic level of an enterprise. Naturally, a vision describes the enterprise's view of its future, which implies a long decision horizon. The RTE was introduced in Sect. 2.1 as a vision that can be applied to manufacturing enterprises, as elucidated in Sect. 2.1. Usually, the enter-

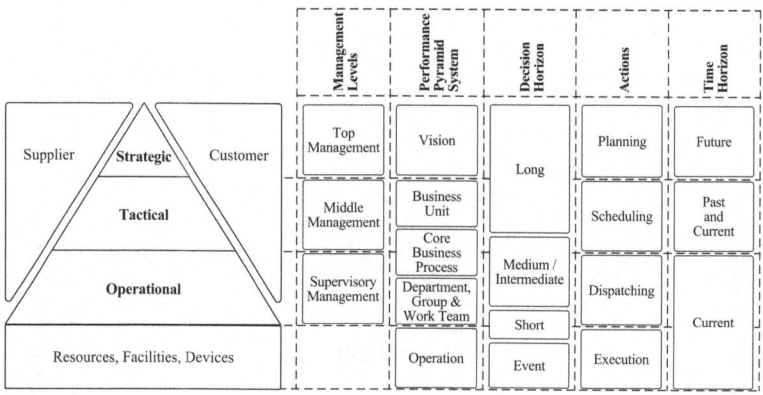

Fig. 3.4 Management perspective of a manufacturing enterprise.

prise strategy is broken down into sub-strategies, of which the *production strategy* is of paramount importance for a manufacturing enterprise [106]. Günther and Tempelmeier describe the indispensable integration of market strategy with production strategy as a sequence of the following steps (cf. [106, 116]):

1. Formulation of *enterprise objectives* with regard to revenue, growth, profitability, and the like, and enunciation of the *enterprise vision* (e.g., RTE).
2. The description of the *market strategy* encompasses a segmentation of target markets, definition of guidelines for product portfolios, deliberations concerning the intended sales volume, and so forth.
3. *Product policy* can be interpreted as a consolidation of the market strategy. The *competitive factors*, like product pricing, product design, product quality, or product service, have to be worked out.
4. Based on the aforementioned product policy, *basic conditions* of production processes are specified, i.e., production methods, material flow, level of automation,

and so forth. This specification of production processes can be performed referring to the scheme depicted in Fig. 3.2.

5. Finally, the *infrastructure of the production system* is envisaged with respect to the above mentioned product policy and the guidelines of production processes. That encompasses the plant layout, quality management, and human resource development, among others.

The last step of the previous sequence (i.e., infrastructure of the production system) can be understood as a transition from top management to *middle management*. At the tactical enterprise level, decisions toward the implementation of the intended strategic goals are made. The questions pertaining to reorganization and further development of the infrastructure of the production system have to be answered [106]. Also, the *control of cost and profit* of the enterprise is located at the tactical enterprise level [101]. Cost accounting, performance measurement (i.e., calculation of KPIs), cost-utility analysis, and the like, are common means to analyze the cost effectiveness of the production factors [205] (cf. also the work of Karadgi [136]). As such, the focus of the tactical enterprise level is about controlling present and past activities of the enterprise.

At the operational enterprise level, decisions concerning the current situation of the manufacturing enterprise are made by *supervisory management*. The operational decisions are the interface between allocation of manufacturing resources and the actual manufacturing process execution [106]. The operational decisions with regard to production planning and control (PPC) are performed at this enterprise level (cf. [70, 106]).

In most instances, PPC is based on the manufacturing resource planning concept (MRP II) encompassing master production schedule, material requirements planning (MRP), capacity planning, (re-) scheduling, and dispatching (cf. [104, 204])[6]. Since the late 1990s, these functionalities have become central for ERP systems that integrate additional departments, like procurement, sales, and logistics.

In contrast to the view that PPC, especially MRP II, is assigned to operational production management (cf. [70, 106]), Higgins et al. present different levels (i.e., strategic, tactical and operational) in the MRP II system [115]. According to this classification, detailed (re-) scheduling, dispatching and shop floor control are interpreted as operational decisions. However, Kurbel concludes that the control aspect of PPC, and thus also the operational production management, is insufficiently supported in today's ERP systems [152]. These deficiencies have led to the development of MES as an intermediate layer between the manufacturing facilities (e.g., resources, machines) and overlying enterprise levels. Also, ERP vendors have acquired MES vendors to fill the aforementioned intermediate layer.

---

[6] In the context of this research work, *push-based production planning and control* is a focus of consideration as it still is the prevalent method in most manufacturing enterprises. Nevertheless, it should be mentioned that *pull-based production planning and control* (like KANBAN) has been developed and is used with certain operational assumptions.

3.3 Engineering Perspective

The main shortfall of the above management perspective results from the missing
feedback between production planning and the actual manufacturing process execution. Consequently, the engineering community has introduced MES to bridge the
vertical integration gap among different enterprise levels. This integration is primarily envisaged to establish multiple closed-loop controls to achieve higher process
efficiencies–concerning quality, maintenance, inventory, and production. Before
delving into different enterprise levels, which have been defined in the engineering
perspective, system theoretical principles with regard to feedback loops are elucidated.

3.3.1 Principles of Feedback Control

The idea of controlling a system is well known in everyday life as it can be described
as an act of producing a desired result [244]. Similarly, *cybernetics* that have mainly
been constituted by Norbert Wiener (cf. [269]) describes (feedback) control within
machines, creatures, and organizations. As already mentioned in Sect. 2.1, the idea
of multiple closed-loop controls has been applied in ISR, and particularly, can be
seen as a foundation for the realization of the RTE (cf. [75, 269]). Furthermore,
information feedback has already been mentioned by Buffa as part of a production
system [36]. Henceforth, the functioning of controlling a system or process is elaborated in the following paragraphs.

In *control engineering* two basic types of control systems are distinguished–
open-loop control and *closed-loop control*, as illustrated in Fig. 3.5. In general,
a control system is comprised of a system (also called plant) or process to be controlled and a controller that exercises control over this plant/process [244]. In the
case of open-loop control, a desired value is given as a set point/reference to the
controller, which in turn derives a manipulated variable as an input for the plant.
However, disturbances also influence the plant, thus affecting the output of the plant
(i.e., controlled variable). Consequently, any error has to be detected *manually* and
the set point has to be adjusted *manually* to compensate for the error that is induced
by the disturbance. In contrast, *closed-loop control* has introduced a capability to
automatically compensate for the disturbance in the plant. The output of the plant is
measured (i.e., controlled variable) and a *feedback* signal/value is sent to a comparator. In most instances, the comparator calculates the difference between the desired
value (set point) and the feedback (i.e., negative feedback). This difference is called
an error and is fed into the controller that calculates a manipulated variable. The
calculation of the manipulated variable is done in such a way that it, usually, will
reduce future errors. Thus, it is set to achieve a state of balance or equilibrium [120].

Summarizing, Mandal defines feedback control (i.e., closed-loop control) as "an
operation, which, in the presence of disturbing forces, tends to reduce the difference
between the actual state of a system and an arbitrarily varied desired state of the

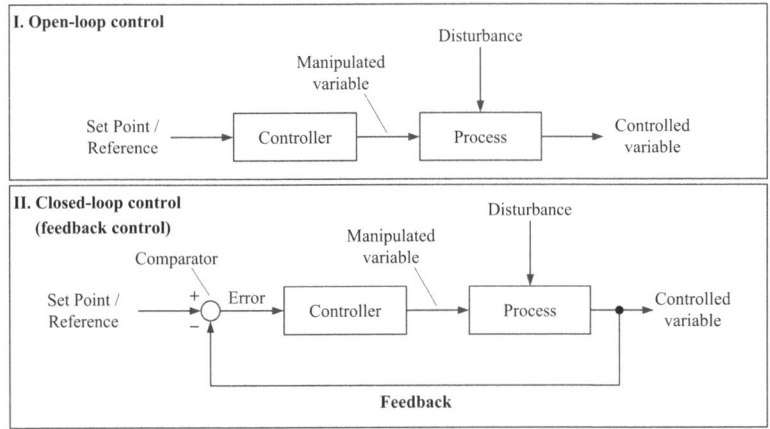

Fig. 3.5 Block diagrams showing cause-effect relationships among different types of control systems (adapted from [166]).

system and which does so on the basis of this difference" [166, 5]. However, a common problem with feedback control is that control actions are only taken *if* an error has already occurred [105]. Hence, *feedforward control* senses the disturbance and tries to anticipate the effect of this disturbance *before* it affects the plant (i.e., proactive control) [105]. Obviously, this requires having a model of the system (plant) to be controlled, especially, its cause-and-effect relationships.

3.3.2 Multiple Control Loops across a Manufacturing Enterprise

The principles of the aforementioned system control–especially closed-loop and feed forward control–can be applied to different levels of a manufacturing enterprise (e.g., [95]). The application of multiple closed-loop controls across different enterprise levels is depicted in the rightmost part of Fig. 3.6. Here, objectives are set at a higher enterprise level and are used as reference for subjacent enterprise level(s). The other way around, decisions on a certain (higher) enterprise level are performed taking the feedback of the subjacent enterprise levels into account. Usually, the aim is to achieve a balanced state (i.e., equilibrium) around the desired values (objectives).

The enterprise pyramid in Fig. 3.6 has similarities to the one presented for the management perspective (cf. Fig. 3.4). However, a vertical integration gap has been identified between the enterprise control level (i.e., often ERP system level) and the actual manufacturing level (i.e., manufacturing process execution) (cf. [247, 255]). This vertical integration gap primarily constitutes divergent time horizons of decisions/reactions and various semantics of information at different enterprise levels.

The (IT) systems and methodologies, like MRP II and later ERP, which have been considered as sufficient means in the management perspective (cf. Sect. 3.2), do not exhibit the required process vicinity for an expeditious and flexible handling of manufacturing processes and prompt optimization [255]. As such, the realization of multiple closed-loop controls is hindered or just inadequately reached [146,201].

Hence, the engineering community has carried out research to close the temporal and semantic integration gap. EI is part of enterprise engineering and is composed of a collection of tools and methods to design and maintain the integrated state of an enterprise [187, 201]. Several enterprise reference architectures have been developed in the realm of EI, which guide the design and implementation of an integrated enterprise [99]. By the mid-1990s, the enterprise integration reference architectures CIMOSA [65], PERA [270], ARIS [223] and GRAI/GIM [42] were presented. Each of the reference architectures focuses on certain requirements of an enterprise.

In contrast, the Generalized Enterprise Reference Architecture and Methodology (GERAM) was developed based upon the above mentioned reference architectures, thus tries to cover more enterprise requirements [20]. Later, GERAM was standardized as ISO 15704–Requirements for Enterprise Reference Architecture and Methodologies [129]. However, none of the enterprise reference architectures mentions how to realize them in terms of (information) technologies [99]. Research and

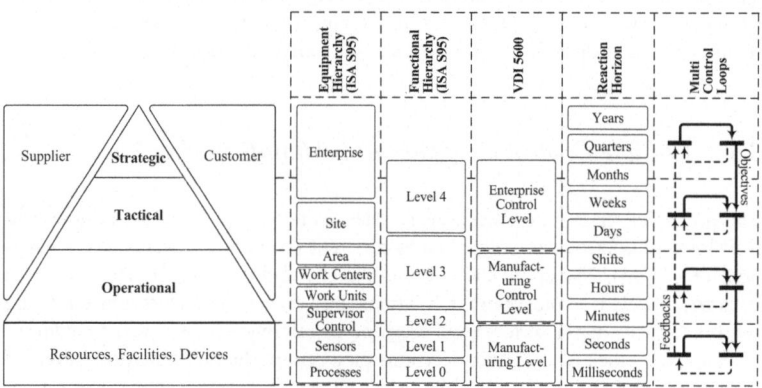

Fig. 3.6 Engineering perspective of a manufacturing enterprise.

development have been performed on MES as a means to implement a vertically integrated manufacturing enterprise. In recent years, some standards pertaining to MES have been issued (cf. [247, 255]). In short, MES tries to bridge the vertical integration gap and target the realization of closed loop controls. Because of this pretention of an online coupling of enterprise planning and actual manufacturing process execution, MES is mentioned as an enabler for the RTE in manufacturing (cf. [168]). Definitions and descriptions of MES have been provided by the Association of German Engineers (VDI), the ISA, and the MESA. Following, the views of VDI, ISA and MESA on MES are summarized.

3.3.2.1 VDI 5600 - Task-oriented View on MES

The VDI intensified its activities on MES around mid-2000s with the foundation of a field of competence Information Technology, and issued the guideline 5600 to provide a task- or problem-oriented description of MES [255]. As depicted in Fig. 3.6, VDI 5600 considers an enterprise as a hierarchy consisting of enterprise control level, manufacturing control level and manufacturing level.

MES is seen as located at the *manufacturing control level*, thus being a link between the tactical level and the actual manufacturing process execution at the manufacturing level. The need for comprehensive, up-to-date, and complex information with regards to manufacturing as well as the deduction of appropriate responses to certain occurrences, situations, deviations, and events necessitates an MES as "a manufacturing management system working close to the [manufacturing] process" [255, 5].

A central idea of MES is that information about production can be used to propose or generate *reactions* that either (i) inform about divergences between a plan and the actual state; or (ii) directly influence the manufacturing process to mitigate or reduce these divergences[7] [255]. Further, guideline 5600 identifies eight tasks that can be part of an MES. Following, these functionalities are listed with emphasis on the requirements to respond and react on certain situations (cf. [255]):

1. *Detailed Scheduling and Process Control*: The task of detailed scheduling is the derivation of a *production plan (schedule)*, i.e., definition of which production order has to be produced when at which production resources. Higgins et al. describe this functionality in the context of MRP II as *execution planning*, thus differ it from the term master production schedule (MPS) [115]. Similarly, detailed scheduling is sometimes known as *fine planning*. The aim of *process control* is to establish a *closed-loop control* employing the production plan (schedule) and *real-time status information* (i.e., status of production orders) from the manufacturing level. The aim is to detect deviations from the production plan and initiate proper (re-) actions to either adapt the production plan or manipulate the actual process execution.

2. *Equipment Management*: Equipment management has to ensure the availability and reliability of machinery, tools, numerical control (NC) programs, and the like. Here, the acquisition and management of current states of resources (e.g., machinery) is seen as a prerequisite for feedback control of production processes.

---

[7] Even though VDI 5600 mentions the necessity of immediate reactions, it does not provide any (information) technologies to achieve this vision.

3. *Material Management*: Material management realizes a timely supply and disposal of materials in accordance with the production plan (schedule). It entails the in-process-control of material qualities and the management of batches.
4. *Personnel Management*: Personnel management aims at *allocation of qualified personnel* for the production process, thus is an indispensable part interacting with detailed scheduling.
5. *Data Acquisition*: The collection, filtering and pre-processing of process data–especially from the manufacturing level–is performed by *event-driven data acquisition*. It is a mandatory basis for most other MES tasks.
6. *Performance Analysis*: Performance analysis is described as establishing *closed-loop controls* with (i) short cycle times (i.e., hour/shift); and (ii) long cycle times (i.e., weeks, month, years). Here, the performance is expressed by means of KPIs, like lead time, availability, and so forth. At the operational level, target performance is compared with actual performance, and (re-) actions are triggered to mitigate occurred deviations. At a tactical level, the trend of KPIs is analyzed and used in the *continual improvement processes* (CIP).
7. *Quality Management*: Quality management aims at sustaining or improving the *product and process qualities*. In the case of detection of quality problems (e.g., non-adherence to product specifications), corrective measures can be taken to minimize (potential) negative effects on the production process.
8. *Information Management*: The main task of information management is the provision of *integration capabilities*. The integration of planned data from enterprise control level (e.g., order details) and actual process data from manufacturing level has to be achieved with information management. Further, it is mentioned that rules can be defined to recognize "fragmentary work steps or extraordinary process states" [255, 42].

According to a study performed by the Fraunhofer Institute for Manufacturing Engineering and Automation (IPA), Germany, more than 60 % of the participants out of 140 manufacturing enterprises have seen the functionalities data acquisition, detailed scheduling, and performance analysis as significant (cf. [271]). In contrast, personnel management and quality management are perceived as less important (cf. [271]).

VDI 5600 provides a comprehensive overview of fundamental functionalities of MES. Other parts of VDI 5600 concern the efficiency of MES (cf. [258]), specify a logic interface for machine and plant control (cf. [256]), and investigate the support of various production systems (e.g., constant work in progress (CONWIP), just in sequence (JIS), Poka Yoke) by MES (cf. [257]). However, VDI 5600 isn't as technical as other standards, like ISA 95. Also, there has been criticism that VDI 5600 is significantly influenced by the Fraunhofer IPA, Germany [159].

3.3.2.2 MESA

The MESA was one of the first not-for-profit organizations that worked on MES. In their initial MESA model eleven functions or activities were described that make up MES [168]. It had been recognized that available MRP II solutions could not provide (near) real-time information about current manufacturing processes [168]. Following, the latest evolution of the MESA model is presented in Fig. 3.7.

Interestingly, the MESA model describes the necessity for a *real-time event environment* that is capable to capture events raised during manufacturing/production. These events can be processed (e.g., collected, analyzed, managed, and reported) at the manufacturing/production operations level, and can be used in several production management functions[8], like performance analysis, quality management, and so forth. The (real-time) monitoring of resource availability, trend monitoring of production performance, and the automatic provision of decision support to operators for improvement or correction of manufacturing process activities are mentioned in the MESA model (cf. [168]). Moreover, the expedience of appropriate and automatic responses to *unanticipated events*, like resource breakdowns, is emphasized [168].

The MESA model in Fig. 3.7 outlines the comparison/balancing of objectives and results at the manufacturing/production operations level. The balancing of the planned situation with the actual situation can be perceived as a managerial closed loop control. The highest level, i.e., strategic initiatives level, has been added to the MESA model to address the "newest element of the enterprise information technology system, the use of real-time information to support broader enterprise objectives" [168, 14]. For instance, the RTE is seen as such an initiative that could be implemented by MES. However, the MESA model doesn't describe (information) technologies that can be used for a proper realization of an RTE initiative, i.e., technical details are out of the scope of the MESA model. Also, the MESA neither elaborates on interfaces and interactions among various MES functions nor elaborates on the interplay of MES with other enterprises' information systems, like ERP systems.

3.3.2.3 ISA 95

ISA 95 is a standard for the interface between the enterprise control and the actual manufacturing processes. The standard has been transferred to IEC 62264 for its international dissemination and acceptance. Hence, the following discussion on ISA 95 is based on IEC 62264, which is known as enterprise-control system integration. IEC 62264 is based on the Purdue Reference Model (PRM) for Computer Integrated Manufacturing (CIM) and represents a partial model as defined in ISO 15704 [126].

In IEC 62264-1: Models and Terminology, an enterprise is considered as being composed of several hierarchical levels. At each level certain functions are executed, and information is processed and exchanged between adjacent levels. Therefore, a functional hierarchy model has been developed, as depicted in Fig. 3.8. The presented hierarchy can also be interpreted as a control hierarchy where decisions have to be made at each level. IEC 62264 describes the functionalities at each level

---

[8] Here, the functions are not explained in detail as they are more or less similar to the MES functions mentioned for VDI 5600.

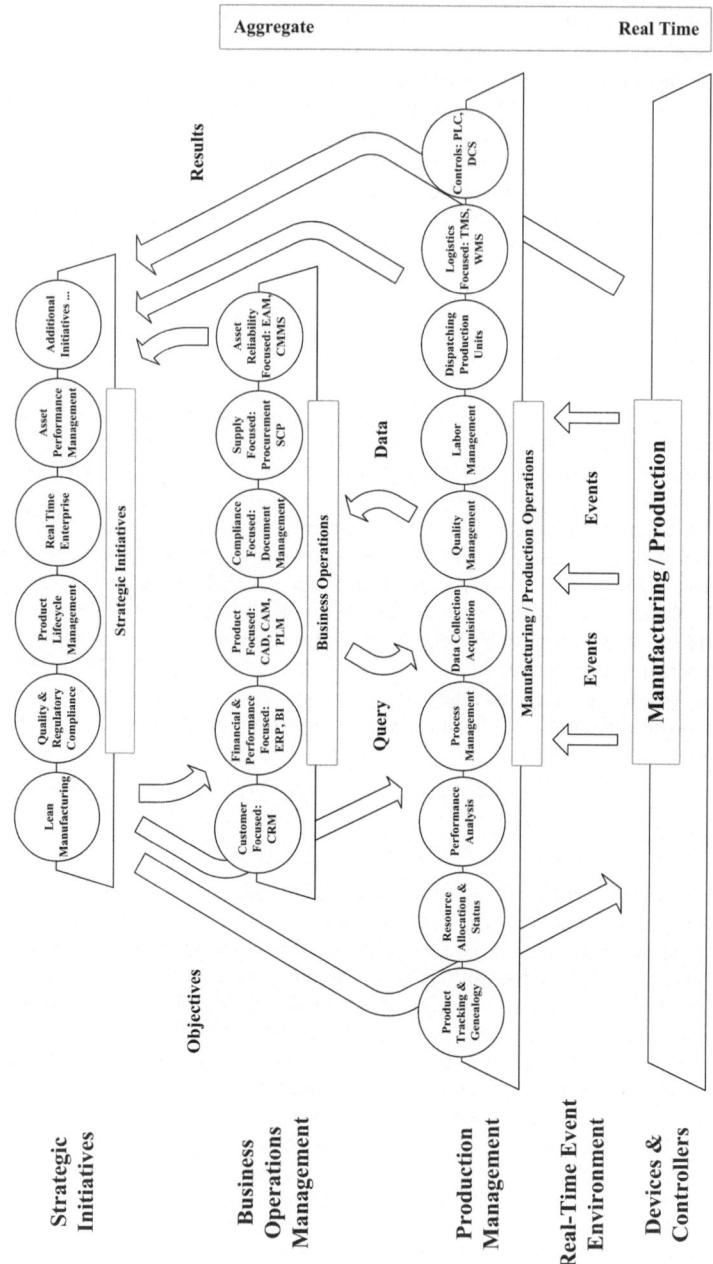

Fig. 3.7 MESA model highlighting vertical integration of enterprise levels employing shop floor events, and envisioning of control cycles between enterprise levels (adapted from [168]).

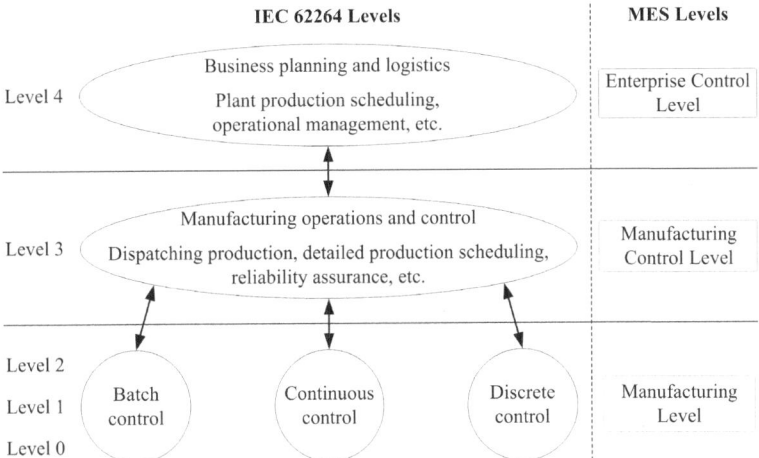

Fig. 3.8 Functional hierarchy model as defined in ISA 95/IEC 62264 [126], and a mapping to MES levels as described in VDI 5600 [255].

independent of any specific control type (i.e., batch control, continuous control, and discrete control). The levels are described as follows (cf. [127]):

- Level 0 describes the *actual physical processes*, which are manufacturing processes in the context of this research work.
- Level 1 is about activities for *sensing and manipulating* the physical processes, thus operates in (milli-) seconds. Typically, this level encompasses sensors and actuators known for automation systems.
- Level 2 activities *monitor and control* the physical processes in hours, minutes, seconds or even faster. The aim of level 0 - level 2 is to bring/maintain the process to/in stable condition.
- Level 3 activities operate on time frames of days, shifts, hours, minutes and seconds, and make up the *MOM*[9] (e.g., detailed scheduling, dispatching production).
- Level 4 is concerned with *business planning and logistics*, and is performed on time frames of months, weeks and days.

Alongside these levels, IEC 62264 also defines an (physical) equipment hierarchy, as illustrated in Fig. 3.6.

The focus of IEC 62264 is on the interface between the level 3 and level 4 functionalities. Eleven functions of the control domain for level 3 (i.e., the MOM or MES level) are defined (cf. [126]), which are based on the initial MESA model (cf. Fig. 3.7):

---

[9] In the context of this research work MOM and MES can be used interchangeably.

1. *Resource allocation and control* is about the management of resources including machines, tools, labor skills, and the like. It especially encompasses the reservation of resources in accordance with the production schedule. Further, resources' status is gathered in real-time and detailed history of resource usage is provided.

2. Based on prescribed production schedules, *dispatching of production* is responsible for management of production flow with regard to jobs, production orders, batches, lots, and so on. Here, it is also envisaged that the production schedule can be adapted within agreed limits and in (near) real-time in case of the occurrence of certain events (e.g., resource breakdown).

3. The *collection and acquisition of data* concerning production equipment and manufacturing processes is a prerequisite for other functionalities.

4. The control of product quality by a (near) real-time provision of quality data (measurements) gathered from manufacturing and analysis is part of *quality management*. Quality management may derive (re-) actions to mitigate quality problems and ensure adherence of manufactured products to corresponding product specifications. The functionality often includes statistical process control (SPC) or statistical quality control (SQC) (cf. also [105]).

5. *Process management* monitors manufacturing processes and automatically corrects or improves in-process functions. The corrected or improved in-process functions can be (i) *intra-operational*, thus are focused on certain machines or equipment; or (ii) *inter-operational*, and thus address the overall manufacturing process. Process management encompasses alarm management to inform personnel about process deviations.

6. *Product tracking* provides status information about personnel, materials, production conditions, alarms, and the like, that are related to a product. Forward and backward traceability are available if the aforementioned status information is recorded for a product, which can be retrieved at a later point in time.

7. *Performance analysis* provides up-to-the-minute information about resource utilization, actual cycle times, conformance to production schedule, and the like. The actual performance values can also be contrasted with past and expected performance values.

8. The *operations and detailed scheduling* functionality is in charge of deriving a sequence of (production) orders based on priorities, characteristics and production rules. Also, the sequencing considers actual and finite resource capacities.

9. *Document control* encompasses the control of documents, like recipes, drawings, and shift-to-shift communication, and so forth.

10. The provision of up-to-the-minute status information about personnel as well as attendance reporting, certification tracking, and so on is part of *labor management*.

11. *Maintenance management* ensures the functioning and availability of production resources (e.g., machines, tools). It encompasses planned (i.e., periodic or preventive) and urgent maintenance activities.

In addition to these functions, IEC 62264 defines a MOM (or MES) model, as depicted in Fig. 3.9. The yellow colored and wide framed area encompasses the aforementioned eleven functions of level three. In most instances, the functions outside of this area, like order processing, are located at level four and are performed by an ERP system. The exchange of information between these functions–especially between level three and level four–is elaborated in IEC 62264-2. Further, the MES functions of level three have been grouped into (i) production operations management; (ii) quality operations management; (iii) maintenance operations management; and (iv) inventory operations management, as illustrated in Fig. 3.9. These groups also define the main subject areas for control activities of MES, i.e., production, quality, maintenance, and inventory. As already mentioned, IEC 62264 is

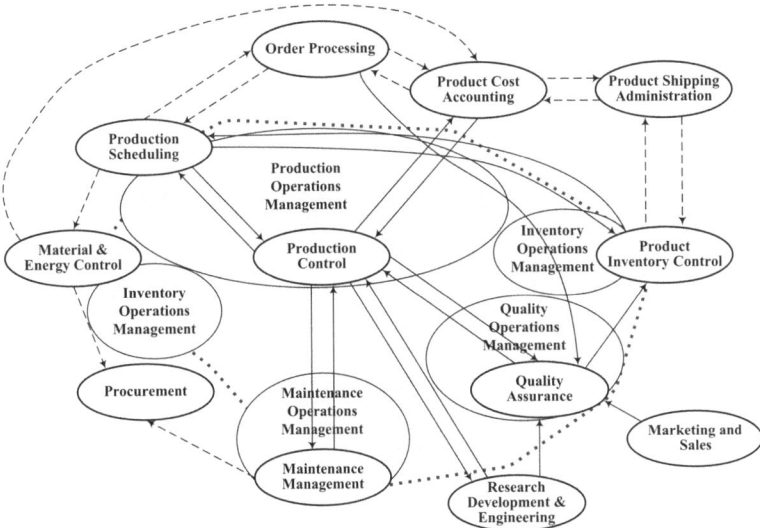

Fig. 3.9 MOM (or MES) model as defined in IEC 62264-3 (cf. [127]).

based on the PRM (cf. also [159]) that was introduced for CIM. The (physical) equipment hierarchy in Fig. 3.6 has been mainly adopted from PRM. As commonly known, most promises made in the context of CIM could not be fulfilled because of an underestimated complexity of challenges that came along with CIM (cf. [224]). Also, innovative IT that could be used for an implementation of CIM was not available.

3.4 Computer Science Perspective

Summarizing the previous discussion on the management and engineering perspectives of a manufacturing enterprise, mainly two challenges can be identified: (i) vertical integration of a manufacturing enterprise (i.e., from the strategic level down to the actual execution of manufacturing processes); and (ii) realization of multiple closed loop controls, which can be performed in (near) real-time.

As IT can provide a significant contribution to the realization of these challenges, the perspective of computer science on a manufacturing enterprise–especially with a focus on the vertical integration and realization of closed loop controls–is depicted in Fig. 3.10. At each enterprise level different information systems have been implemented: (i) ERP systems at the tactical level; (ii) MES at the operational level; and (iii) machines and automation systems at the shop floor. However, the separation of ERP and MES is not always that clear as shown in Fig. 3.10. Rather, it can vary depending on the implemented IT infrastructure within the manufacturing enterprise. Further, the heterogeneity of systems increases in the vertical direction and comes

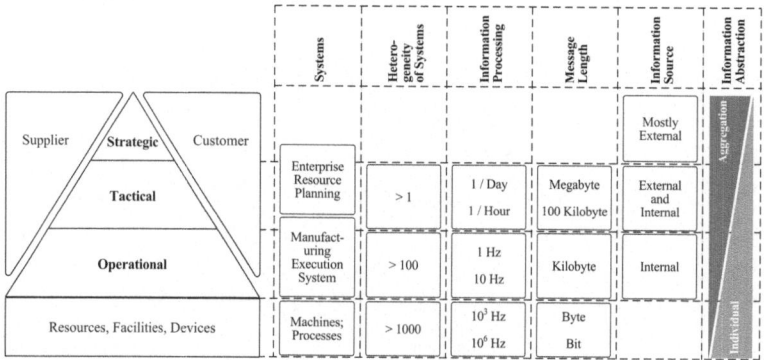

Fig. 3.10 Computer science perspective on a manufacturing enterprise.

closer to physical resources. Unfortunately, this heterogeneity of resources implies significant efforts for integration of machines and automation systems. To overcome this issue, recent endeavors have been made to standardize interfaces between MES and physical resources. For instance, VDI 5600 guideline part three specifies a logic interface for machine and plant control [256]. Recently, under the guidance of the OPC Foundation, the combination of OPC UA and PLCopen (i.e., IEC 61131-3) promises an enhanced interoperability between MES and machines/automation systems (cf. [198]).

The processing of information[10] at the shop floor is characterized by a huge amount of data that is generated in seconds or milliseconds during execution of manufacturing processes. In contrast, at the tactical level, information (data) is processed just in hours or days. In the same way, the individual message (or event) size is much smaller at the shop floor level (e.g., bits up to few bytes) compared to the tactical enterprise level (kilobytes, megabytes). Usually, data or information will

---

[10] In the context of this research work processed data, information can also be interpreted as events or at least as part of events (cf. Sect. 2.2).

be aggregated at higher enterprise levels, whereas individual data or information is most often processed at lower enterprise levels. Further, business applications at the tactical enterprise level are described as transaction-based while the automation systems at the manufacturing level are seen as event-based [146]. Overall, a temporal and semantic integration gap can be identified between the enterprise levels (e.g., [137]).

Considering the requirements of the management and engineering perspectives (i.e., vertical enterprise integration and multiple closed loop controls) in light of the aforementioned circumstances, some implications/requirements for the intended (controlling) IT (system) can be described:

- Events generated at the shop floor have to be processed in *real-time*. Depending on the strictness of deadlines for processing, Kopetz distinguishes (i) soft real-time (online)[11]; (ii) firm real-time; and (iii) hard real-time [148]. If a deadline for processing of an event is missed, but the derived result is still useful, the system is called a *soft real-time system*. If a derived result is useless after violation of the deadline, the system is a *firm real-time system*. Finally, a *hard real-time system* is one in which the violation of a deadline can cause severe problems (e.g., damage of physical equipment), thus all deadlines have to be strictly kept. As the intended controls are planned at higher abstraction levels (e.g., manufacturing control level, cf. Sect. 3.3.2.1) only soft or firm real-time deadlines are considered. Technically, this requires a processing of events with low latencies (cf. also the discussion on latency in Sect. 2.1). Likewise, the control of PLCs/devices at the manufacturing level needs to adhere to hard real-time.
- The events from the manufacturing level have to be analyzed and interpreted incorporating their technical and business context (cf. the control loops in Fig. 3.6). The aim is to assign events to their respective enterprise process entities, like production orders, product specifications, and so forth. Thereby, comparisons between a planned situation and actual status (e.g., concerning product quality) shall be performed. Also, the control should be aligned, for instance, with certain product specifications, thus considering the exact context information.
- Critical situations cannot always be detected considering and analyzing simple events. Rather, these situations are often affiliated with complex events, i.e., patterns of events (cf. Sect. 2.2.3 for a broader discussion of complex events). Hence, a capability to detect and react on those *complex* events is indispensable. In this context, it has to be mentioned that there is a trade-off between the ability to handle/detect complex event patterns and the necessity to react on critical situations in (near) real time (cf. [178]).
- In addition, the need for flexibility and agility has been discussed for the RTE in Sect. 2.1. In the context of manufacturing enterprises, these requirements become apparent as enterprises are forced to follow a high product variety and low production volume (cf. Sect. 3.1.1). Consequently, the manufacturing systems (e.g., machines) and the manufacturing control system have to reflect this

---

[11] The terms soft real-time (online) and near real-time can be used interchangeable throughout this research work.

flexibility and agility, which requires mechanisms to conveniently adapt the manufacturing control system. Ideally, the event processing logic has to be separated from the application logic, avoiding a rebuild of the manufacturing control system (cf. Sect. 2.2.6.3).

3.5 Discussion of Perspectives

The above mentioned presentation of management, engineering and computer science perspectives on a manufacturing enterprise can be summarized as follows. At the strategic enterprise level, for instance, the vision of the RTE has been elaborated (cf. also Sect. 2.1). Starting from such strategic considerations, objectives are broken down into subjacent levels. The operational enterprise level is not omitted, but neglected in the sense that (push-based) approaches, like MRP II, are seen as sufficient means for control of manufacturing. Advanced IT concepts toward a vertical integration of manufacturing enterprises are missing.

In contrast, the engineering community has focused primarily on the actual manufacturing processes. Typically, control engineering has provided concepts for control (i.e., open loop and closed loop control) of technical systems and processes (cf. Sect. 3.3.1). The control of certain machines, robots and automation systems at the manufacturing level has been researched in depth. Starting from the shop floor level, research on MES has been carried out to bridge the vertical integration gap (cf. [126, 174, 255]). Here, functionalities of MES and their interplay have been defined.

Despite this progress, certain issues still remain open with respect to the interface between the enterprise control level and the manufacturing level [201]. In practice, the exchange of data is done manually or at most semi-automatically because of inflexible and proprietary interfaces [140]. Recently, attempts have been made to address the issues by standardization activities initiated by the Association of German Engineers (cf. [256]) or the OPC Foundation (cf. [198]).

From computer science perspective, paradigms, concepts and technologies have been provided for a horizontal integration of an enterprise [94]. EAI based on SOA has been evolved as a de facto standard for interoperability of enterprise applications, like ERP, CRM and SCM systems. Recently, attempts have been made to vertically integrate manufacturing enterprises based on EDA and CEP (e.g., [96, 101, 138]). CEP can especially be considered as a promising candidate for the implementation of an integrated manufacturing enterprise and the realization of multiple closed loop controls at the manufacturing control level. Even though the necessity for event processing has been described, for instance, by the MESA (cf. Fig. 3.7), the integration of MES and CEP requires more attention in research.

Chapter 4
Event-Driven Framework for Real-Time Enterprise

The previous discussion on the realization of RTE for manufacturing enterprises–also considering managerial, engineering and computer science perspectives–reveals the urgent need for (i) vertical integration of various enterprise levels (i.e., from shop floor to top floor); and (ii) realization of multiple control loops between these enterprise levels. Consequently, the aim of the presented research work is the development of an IT-framework for real-time monitoring and control of manufacturing processes. Thereby, previous achievements that have been made by management, engineering and computer science communities are incorporated.

4.1 Literature Review

As the focus of this research work is on manufacturing process control aspects within RTE, recent research activities and developments pertaining to intelligent manufacturing control are elaborated in subsequent paragraphs[1].

4.1.1 Architectural Styles for Manufacturing Control

Diltis et al. have described four basic architectural styles for manufacturing control systems: (i) centralized; (ii) hierarchical; (iii) modified hierarchical; and (iv) heterarchical control [55]. A *centralized control* approach is characterized by a single control system that is in charge of controlling certain process units, like products, materials, orders, and so forth [55]. In practice, the control task has been divided into several control levels (cf. Chap. 3), thus *hierarchical control* architectures have been developed. This hierarchical structure of (enterprise) processes is indis-

---

[1] Some practical and pragmatic aspects and approaches for manufacturing control have already been outlined in Chap. 3.

pensable for reduction of control complexity (cf. [272]). Therefore, it is still the prevalent control architecture in industry.

Nevertheless, Leitao mentions problems of hierarchical control systems to properly (re-) act on disturbances during manufacturing execution [157]. Therefore, decentralized or even autonomous control approaches have been envisioned where physical systems (e.g., machines) and control systems converge (cf. [226]), as depicted in Fig. 4.1.

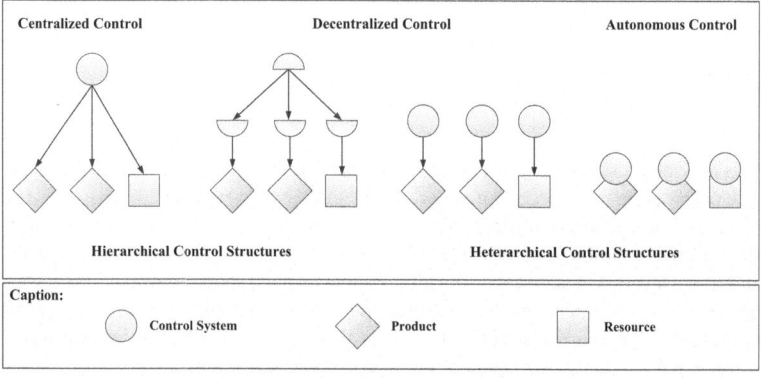

Fig. 4.1 Control architectures and convergence of control and physical systems (adapted from [226]).

4.1.2 Research on Intelligent and Decentralized Control Approaches

Multi-agent systems (MAS) and holonic manufacturing systems (HMS) are the most prominent developments in the realm of decentralized and self-organizing control approaches. According to MAS and HMS, a complex control system can be split into decentralized control units, thus reducing the overall complexity of the manufacturing control task [263]. Instead of making control decisions in a central system that tries to cover all aspects of a certain manufacturing system, decisions can be made autonomously by decentralized control units. Further, decisions are made at the POA incorporating information from the POC (cf. Sect. 2.1). Each control unit is autonomous in the sense that it tries to follow its own objectives and reach a local optimum [157]. Consequently, distributed control units need to communicate with each other to achieve objectives with regard to the overall (manufacturing) system.

4.1.2.1 Multi-Agent Systems

MAS can be employed for the realization of the above mentioned distributed control principles in manufacturing. Here, an agent is an "autonomous component that represents physical or logical objects in the system, capable to act in order to achieve its goals, and being able to interact with other agents, when it does not possess knowledge and skills to reach alone its objectives" [157, 982]. An agent realizes the concepts of autonomy, encapsulation, goal-orientation, interaction, persistence, and reactivity [263].

Approaches to implement intelligent decision-making capabilities and reasoning within (software) agents are primarily taken from artificial intelligence (AI) and operations research (OR). However, the use of AI and OR methods in the realm of complex manufacturing systems is still in its infancy [189]. MAS establishes a heterarchical control architecture where control units are considered as being equivalent and not subordinated to each other. Conceptually, distributed control approaches, like MAS, possess the advantage of being more flexible and robust to disturbances, but also imply unpredictable control behavior [263].

Sundermeyer and Bussmann (cf. [241]) are recognized as probably the first authors who have reported on a full-scale *industrial* agent-based production system that has been adopted by a manufacturing enterprise [157]. This MAS has been implemented by Schneider Electric GmbH, Germany, and has been used by the Daimler AG (- formerly DaimlerChrysler AG -), Germany, to flexibly control a transfer line for the production of cylinder heads [47]. Hence, this MAS was intended to optimize the material flow of a large-volume production [241].

In addition to the aforementioned MAS of Sundermeyer and Bussmann (cf. [241]), a framework based on MAS technology and aimed at realization of autonomous and cooperative control of complex production systems has been proposed by Mönch and discussed by the ISR community [189]. Mönch advocates the thesis that a renunciation of any form of control hierarchy does not offer better results than a priority-based approach [189].

Consequently, the proposed framework is described as an instrument for the realization of a *distributed hierarchical control*, thus can be considered as being of a hybrid nature. Overlying control levels define a target corridor for subjacent control levels, wherein system units perform autonomous decision-making processes. Even though the framework has been *conceptually* implemented, its industrial implementation and adoption are still pending. Similar to Mönch, Scholz-Reiter and Freitag argue that the future shift from *offline* planning to *online* control of production processes requires additional distributed control units *beneath* established control levels, thus sustaining ERP systems [226].

In addition to MAS, several flexible manufacturing control approaches have been developed within the intelligent manufacturing systems (IMS) program (cf. [122]). These approaches have been inspired by nature, mathematics, and social organizations [157], and are elaborated in subsequent paragraphs.

4.1.2.2 Bionic Manufacturing Systems

In a bionic manufacturing system (BMS), production units are represented as cells of a living organism [246]. The stability of a living organism is maintained by regulation of metabolic reactions through enzymes and hormones, which can be translated into coordinators and policies/strategies for production systems [246]. Interestingly, the cells (production units) are organized in a *hierarchical* manner and urgent reactions are initiated by a *central* nervous system (i.e., central control system) [246]. Hence, the analysis of living organisms and their control reveals the necessity for a *central controller*.

4.1.2.3 Fractal Manufacturing Systems

Fractal is a mathematical concept for the description of recursive and self-similar patterns. Fractal is a word of Latin derivation meaning broken or fractured (into smaller pieces). Similarly, a fractal manufacturing system can be described as composed of fractal entities [246]. A fractal can represent an entire manufacturing shop at operational level as well as a production resource (e.g., machine) at the shop floor [221]. The fractals possess self-organizing capabilities and can adapt dynamically to environmental influences [246]. Conflicts and disturbances in the manufacturing system are resolved through cooperation and negotiation between fractals, thus requiring efficient information and communication systems [221]. Although a fractal manufacturing system assumes a hierarchical architecture, the created hierarchy is not as rigid as for traditional hierarchical architectures.

In most instances, theoretical issues with regard to fractal manufacturing systems are discussed, but implementation has been neglected [221]. Therefore, Ryu and Jung have presented an agent-based architecture for fractal manufacturing systems, as depicted in Fig. 4.2 [221]. A basic fractal unit (BFU), e.g., representing a machine, has five functional modules: (i) observer as an interface to physical devices (sensors) or other BFUs; (ii) analyzer for evaluation of alternative job plans; (iii) organizer that manages fractal status and fractal addresses for dynamic reconfiguration processes; (iv) resolver for job plan generation, initiation of cooperation and negotiation, and decision making; and (v) reporter as an interface to other BFUs or physical devices (actuators). As such, each BFU realizes a *closed-loop control* following the *sense-analyze-respond* principle (cf. Sects. 2.2.5 and 3.3.1).

4.1.2.4 Holonic Manufacturing Systems

Arthur Koestler has described the hybrid nature of basic units of life in his famous book "The Ghost in the Machine" [147]. He coined the term *holon* composed of the Greek words *holos* (i.e., whole) and *on* (i.e., particle) to denote that everything is a whole and part simultaneously [157]. Koestler also identified two main character-

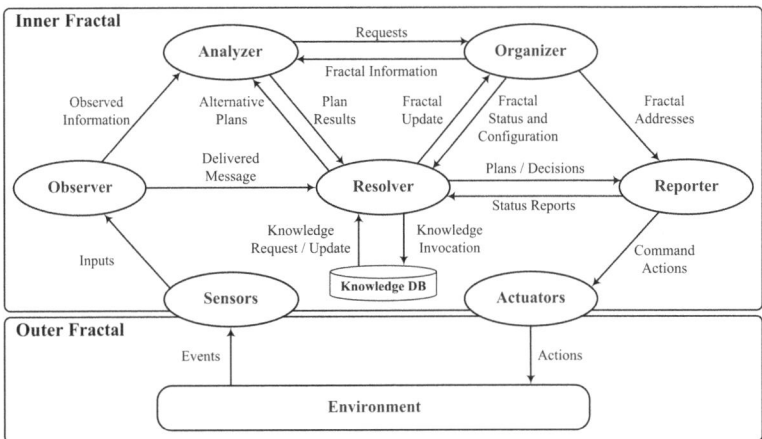

Fig. 4.2 Fractal architecture and relationships among functional modules as part of a basic fractal unit (adapted from [221]).

istics of holons: (i) autonomy as the ability to act autonomously in case of unpredictable circumstances; and (ii) cooperation as the ability to cooperate [157, 246].

The concept of holons can be transferred to manufacturing systems. For instance, a machine is a whole and can control itself, but is also a part of a production line. Hence, holons build a *hierarchical structure* that is called *holarchy.* However, a certain holon can belong to many holarchies at the same time [157]. A reference architecture for HMS called PROSA has been presented, which defines three basic holon types: (i) a *resource holon* for physical parts, like a production resource (e.g., machine); (ii) a *product holon*, which keeps knowledge about the (manufacturing) process and the product; and (iii) an *order holon* representing a certain task in the manufacturing system [33]. As an extension, the staff can also be represented by *staff holons* [33].

Further, a HMS "*must* have a direct interface to the shop floor" [171, 235], where a holon–resource or product holon–encompasses a hardware/equipment part for manipulation of the physical world (e.g., devices, conveyor belt) and a control part responsible for hardware/equipment control and communication with (adjacent) holons [171]. Such *real-time information feedback* from physical resources is seen as a prerequisite for dynamic adaptation of, e.g., assembly systems and processes [226]. Hence, the vision of HMS has influenced standards, like IEC 61499 and IEC 61131-3, which define programming languages and function blocks for PLCs. IEC 61499 deals with the development of distributed control systems based on a function block model [171].

The basic structure of a function block is depicted in Fig. 4.3. The instance of a function block is a software unit that encompasses a copy of a certain data structure and assigned operations for processing (e.g., manipulation) of (input) data [128].

The execution is controlled in an *event-driven* manner, thus function block instances are triggered by input events. Further, function block instances are organized and

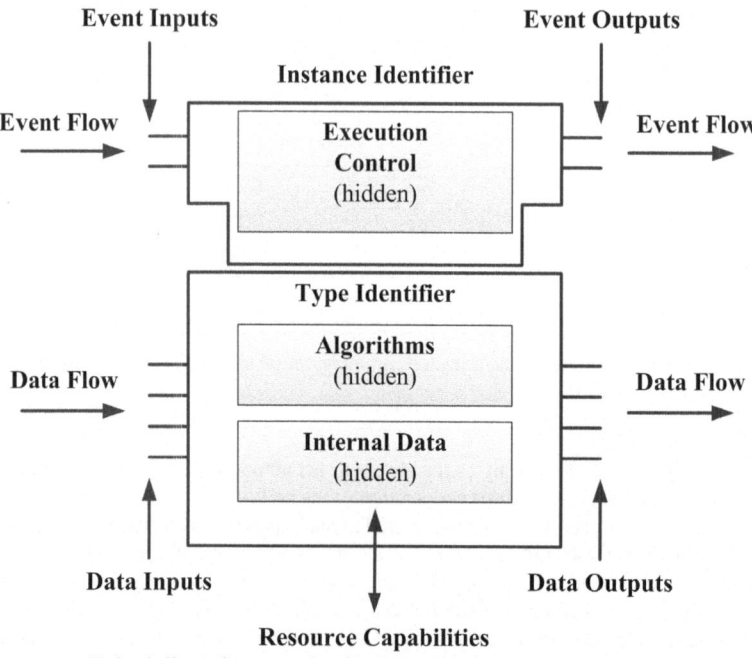

Fig. 4.3 Properties of function blocks according to the function block model of IEC 61499 (adapted from [128]).

orchestrated to function block networks. As such, the execution control according to IEC 61499 has similarities with the *event processing model* presented for EDA and CEP (cf. Sects. 2.2.5 and 2.2.6). However, the event-driven control, as presented in IEC 61499, is a new paradigm that is in its infancy [171]. Thus far, IEC 61499 has not been well adopted by industry [276]. Further, the processed events in IEC 61499 are considered as simple events, thus methodologies for a complex processing of events are more or less out of the scope of IEC 61499.

Overall, MAS focus on implementation and technical aspects and HMS concerns more about conceptual issues. MAS have been used as a technology to implement HMS. Hence, Mařík et al. summarize the differentiation of MAS and HMS as follows: "Holonics is not a new technology, but is a system-wide philosophy for developing, configuring, running and managing a manufacturing business. Multi-agent system area is a promising technology provider to accomplish the ambitious holonic visions" [171, 236]. HMS follow an event-driven control strategy while MAS foster distributed information processing, and both paradigms have been considered as important breakthroughs for intelligent manufacturing control [171]. Consequently, several research projects have been initiated to capitalize on the HMS philosophy and MAS technology.

4.1.3 Service-Orientation for Manufacturing

Complementary aspects of MAS and HMS, like modularity, flexibility, and reconfigurability, are researched and developed in EU-funded programs [158]. Here, a focus on service-oriented concepts in the realm of monitoring and control of (production) processes can be recognized.

Karnouskos et al., for instance, identify an insufficient vertical integration from top floor (i.e., ERP system) to shop floor (i.e., physical resources) primarily because of proprietary interfaces [141]. Consequently, SOA-ready industrial devices are proposed by Karnouskos et al. [141]. These devices can be parts in a (business) process orchestration that is common in almost every SOA. A prototype has been implemented within the SOCRADES project (cf. [236, 237]) as a proof of concept. Web services have been embedded into physical devices of a *laboratorial* production line following the device profile for web services (DPWS) [141]. DPWS is an OASIS standard for implementation of web services at the physical device level and encompasses, for instance, the firing and receipt of *simple events* between devices [58].

The same technological approach has been used in the SIRENA project which aims to develop a service infrastructure for real time embedded networked applications [131]. DPWS has been selected after an evaluation of technologies for device integration, like Universal Plug and Play (UPnP) and JINI [29]. However, real-time requirements, which appear in the realm of manufacturing control, usually demand close coupling, thus making the applicability of the SOA paradigm at the manufacturing level complicated [102].

Nevertheless, Gartner also stressed SOA is an emergent paradigm to integrate manufacturing operations into demand-driven value networks [130]. Their approach has been called *Manufacturing 2.0*, which "leverages service and collaboration based architectures for manufacturing—right first time and on-demand—across dynamically reconfigurable sensor and mobile worker enabled supply networks" [130, 10]. The vision of Manufacturing 2.0 is the achievement of seamless integration among PDM, ERP, MES, and the like. Hence, a suitable reference model has been proposed to guide implementations of Manufacturing 2.0 [130].

4.1.4 Convergence of Control and Physical Systems

In recent years, the vision of an Internet of Things (IoT), i.e., the connection of physical things with the Internet, appeared and has been extensively researched. The advent of emergent technologies, like RFID, has ushered in a convergence of control and physical systems. In the context of RTE, this convergence has been mentioned as a means to establish seamless flows of information from POCs to POAs (cf. Sect. 2.1.1).

Individual physical products have been proposed as additional elements of control in the overall control structure of a production system [274]. Physical products equipped with RFID transponders are considered *smart products*, thus can exchange information with other control levels and resources [274]. Zäh et al. propose to store the complete product-related information *locally* on RFID chips [274]. However, this distributed management of product data as well as the limited available data volume on RFID chips are raising questions concerning product tracking and tracing (cf. MES functionalities mentioned in VDI 5600 and IEC 62264). A prototype of the adaptive, product-based control system has been implemented in a laboratory at the Technical University of Munich (TUM), Germany.

Initiated in 2009 by the National Science Foundation (NSF), USA, research is carried out on cyber-physical systems (CPS). According to the initial definition of the NSF, the term CPS "refers to the tight conjoining of and coordination between computational and physical resources" [192, 4]. A CPS bridges the cyber-world of computing and communications with the physical world [208]. CPSs are "physical and engineered systems whose operations are monitored, coordinated, controlled and integrated by a computing and communication core" [208, 731]. Recently, the future project "Industry 4.0" was initiated by the German Federal Ministry of Education and Research (BMBF) as part of its high-tech strategy 2020. The *vision* of the research program encompasses the development of interconnected CPS in the realm of smart production and smart factory [54]. Consequently, decentralized and autonomous control approaches for manufacturing processes are envisioned that capitalize on future CPS. Currently, research and development on CPS, especially, toward intelligent manufacturing control is in its infancy.

4.1.5 Event Processing in Manufacturing

The (near) real-time reaction to disturbances and unexpected events is recurrently mentioned as a central challenge for intelligent manufacturing control. However, available CEP solutions are primarily focused on business and financial processes. Research and development pertaining to the appliance of CEP for (near) real-time control of manufacturing processes is rare [101, 275].

A system architecture for a unified event management has been conceptualized by Walzer et al. to deal with complex events for monitoring and control of manufacturing processes [265]. The proposed architecture captures events from the manufacturing level, but cannot integrate these events with transactional data from the enterprise control level. Similarly, Zhang et al. present an extensible event-driven manufacturing management with CEP approach [275]. Their approach is based on an MES platform, thus uses both shop floor events and transactional data from an ERP system. However, technical details of the presented MES approach remain unclear (e.g., data collection).

System Insights has developed an open source framework for real-time monitoring and analysis of manufacturing enterprises [261]. The framework consists of data delivery, data collection, and data analysis. CEP engines are employed for data analysis incorporating temporal and causal relations between events. An event-driven framework for real-time monitoring and control of manufacturing processes based on CEP has been developed and implemented [101, 138]. This control system has been implemented for a foundry, thus has been developed and validated in a real production environment.

4.1.6 Summary

As elaborated in the above sections, extensive research has been carried out on decentralized and autonomous control approaches. However, purely decentralized control systems have not been adopted widely in industry. In the context of MAS and HMS, Leitao and Vrba, for instance, state that "the majority of reported applications are found in academic and laboratorial environments and not in real plants" [158, 16]. Similarly, Sundermeyer and Bussmann criticize this by stating that there is nearly no (industrial) proof of the applicability and promised benefits of MAS for production control [241]. Henceforth, most researchers have come to the conclusion that a hierarchical control structure has to be sustained and might be supplemented by decentralized control units (e.g., [189, 226, 246]).

The latest research activities aim to introduce the SOA paradigm throughout all control levels of a manufacturing enterprise (i.e., from top floor to shop floor). In addition, the systematic processing of events that are generated during manufacturing processes have been identified as a promising approach for the realization of RTE in manufacturing [95]. Although CEP is a proven technology in the financial industry, its adoption for real-time monitoring and control of manufacturing processes requires further attention [101].

In the following sections, an event-driven framework for real-time monitoring and control of manufacturing processes is elaborated; it capitalizes on EDA and CEP. This framework realizes an event-driven MES and establishes (i) a vertically integrated manufacturing enterprise; and (ii) multiple closed-loop controls between various control levels. As such, it follows the basic principles of RTE (i.e., integration, automation and individualization) as described in Sect. 2.1.1. The design of the framework respects management, engineering, and computer science views of a manufacturing enterprise and incorporates their relevant concepts, as delineated in

Sects. 3.2, 3.3, and 3.4. Further, a process model that guides the implementation and introduction of the framework in a manufacturing enterprise is presented.

4.2 Process Model toward the Realization of RTE

An overview of the process model toward the realization of RTE in manufacturing is depicted in Fig. 4.4. The description of the process model is mainly adapted from the author's published and awarded[2] work [98, 182]. The process model consists of four process steps, which are unique to a particular manufacturing enterprise. It is not required to perform the process steps in sequential order. Rather, individual steps should be carried out once in a while to improve the value creation processes[3]. As is common, for instance, in any BPM life cycle (cf. [2,266]), the implementation

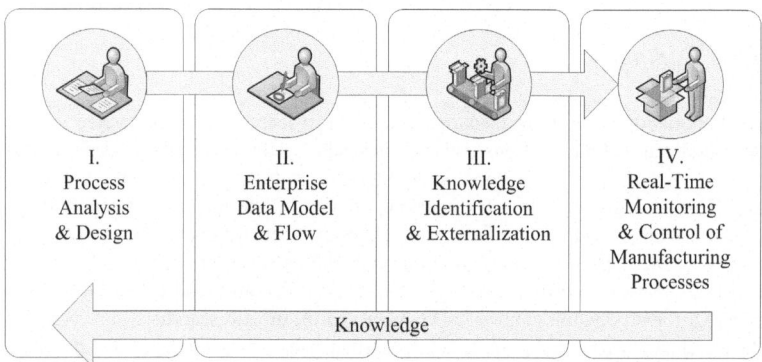

Fig. 4.4 Process model toward the establishment of RTE in manufacturing as published in [96,98, 182].

of the envisioned IT framework is necessarily prepared by analysis and (re-) design of value creation processes (cf. step I in Fig. 4.4). The primary aim of the vertical integration of a manufacturing enterprise is the reconciliation of targets that have been set at the enterprise control level with the actual situation at the manufacturing level. Hence, an enterprise data model based on standards, like IEC 62264 (cf. Sect. 3.3.2.3), is designed to relate planned and actual process values (cf. step II in Fig. 4.4). This enterprise data model can be assisted with data flow diagrams (DFDs) to reveal the interdependencies among business applications and machines, and associated events triggered at different enterprise control levels.

---

[2] The author's work was decorated with a best paper award at the Second International Conference on Information, Process, and Knowledge Management (eKNOW 2010), St. Maarten, Netherlands Antilles.

[3] In the following, value creation processes are primarily understood as manufacturing processes.

The IT framework envisions the persistent storage and management of process data along with associated planned values in a relational database. The database schema is set up according to the above mentioned enterprise data model. The integrated enterprise data contains implicit knowledge about the executed manufacturing processes (e.g., possible reasons for a machine's malfunction). Therefore, an offline knowledge discovery in database (KDD) processes can be employed on the integrated enterprise data to identify new knowledge (cf. step III in Fig. 4.4). Naturally, a methodology for collecting process data[4] is a prerequisite for the execution of the KDD process.

Applying repeated and time-consuming database queries executed on the integrated database are not useful for monitoring and control of value creation processes [38]. Instead, event streams created during manufacturing process execution can be analyzed and processed in (near) real-time employing CEP (cf. step IV in Fig. 4.4). The knowledge that has been externalized in the KDD process can be expressed in terms of event pattern rules. Event patterns describe critical process situations that require appropriate (re-) actions. By deduction of these (re-) actions and their immediate implementation, value creation processes can be improved in the sense that the actual process values are reconciled with their associated planned values. In the following subsections, each of the above process steps will be elaborated.

4.2.1 Analysis and (Re-) Design of Value Creation Processes

Usually, enterprise reference architectures, like GERAM or ARIS (cf. Sect. 3.3.2), and BPR (cf. [108]) encompass analysis and (re-) design of value creation processes. The following phases are carried out in BPR: (i) identification of critical value creation processes; (ii) review, update and analysis of value creation processes (AS-IS analysis); (iii) (re-) design of value creation processes based on AS-IS analysis, and (iv) implementation of (re-) designed value creation processes.

Comprehension of value creation processes and their vertical integration with overlying control levels is crucial for the implementation of real-time monitoring and control approaches. Further, process analysis and (re-) design incorporates the enterprise's organizational structure as well as its process-oriented organization.

A value creation process is composed of several activities/functions that are executed or supported by various resources (e.g., IT systems, machines). Each organizational unit is responsible for a set of process functions. These organizational units are organized and modeled employing *organizational charts*. A process function can take different types of informational inputs (e.g., process plan) as well as implicit knowledge of workers. Overall, an enterprise's process function transforms the inputs to outputs, and an organizational unit is in charge of performing this trans-

---

[4] A detailed description about the collection of process data in (near) real-time is given in Sect. 4.3.1.

formation. Finally, process functions are orchestrated using logical operators, like "and", "or", and "exclusive or", to make up the value creation processes.

A multitude of modeling languages have been proposed to model value creation processes, like EPC, and business process modeling language (BPML). In addition, knowledge intensive processes, i.e., value creation processes that create and expend (implicit) knowledge while executing, can be described using knowledge management description language (KMDL) [103], and the like. Also, BEMN can be employed to model the usage of complex events during the execution of value creation processes (cf. Sect. 2.2.6.4).

4.2.2 Enterprise Data Model and Data Flow Diagrams

Today's manufacturing processes are described by Ovacik and Uzsoy as being complex [200]. According to Mönch, complex manufacturing processes are characterized by (i) fluctuating product mix; (ii) coexistence of different manufacturing types (i.e., batch and discrete manufacturing); (iii) execution of multiple orders on a single machine; (iv) strict delivery times (i.e., for instance, to support KANBAN); (v) high complexity of process plans; (vi) changeover times that depend on production schedule; and (vii) existence of parallel machines [189].

Often, these manufacturing processes are equipped with automation systems and machines. Enormous amounts of process data are generated by automation systems in real-time, denoting information, like process performance, machine breakdowns, product positions, and so forth. The real-time monitoring and control of complex manufacturing processes has to capitalize on this process data. Hence, process data generated by business applications and automation systems has to be analyzed to identify critical control-related process parameters. In this regard, enterprise data modeling is an indispensable step to reveal the interdependencies of process data. It influences the quality of information that is required to execute value creation processes, achieve EI, and enhance real-time monitoring and control of value creation processes.

As described in Sect. 3.3.2.3, IEC 62264 describes models and terminologies that guide the implementation of holistic monitoring and control of manufacturing processes. IEC 62264-2 encompasses a specification with regard to the exchange of process and control data between the enterprise control level and the manufacturing level (cf. Sect. 3.3.2.1). In most instances, an *enterprise data model* in the realm of MES is based on IEC 62264-2[5], and it can be further augmented with technical models depending on the employed manufacturing types (e.g., DIN EN 61512-2 for batch manufacturing).

In addition to the static structure of the enterprise data model, DFDs are used to expose interdependencies among resources (e.g., machines, business systems), either in isolation or in combination [144]. DFDs describe the dynamic behavior

---

[5] Many MES vendors refer to ISA 95, which has been standardized as IEC 62264.

of value creation processes by displaying flow of data among process resources. As such, a DFD reveals the relationships between various IT-systems and resources (e.g., machines). A manufacturing system contains numerous data flows among automation systems that induce a complicated network of system dependencies. To handle this complexity, DFDs can be organized in a hierarchical manner. A coarse-grained DFD can provide an overview of the shop floor and its resources, while other DFDs cover certain manufacturing (sub-) processes.

4.2.3 Knowledge Identification

The relation between data, information and knowledge can be expressed as follows: "Data is raw numbers and facts, information is processed data, and knowledge is authenticated information" [9, 109]. As stated above, process data can represent (implicit) knowledge that is enriched and enlarged during execution of manufacturing processes (e.g., feedback). The process data can be collected from shop floor resources and integrated with transactional data extracted from business applications. This integrated data is managed in a relational database, whose database schema is defined according to the enterprise data model.

Historical data that is acquired during execution of manufacturing processes can be analyzed to derive new knowledge. The knowledge describes, for instance, process disturbances and their possible root causes, thus can be reasonably used to improve real-time monitoring and control of manufacturing processes. A systematic identification of knowledge can be assisted by employing a KDD process that is "one of mapping low-level data into other forms that might be more compact, more abstract, or more useful" [71, 39]. The KDD process is depicted in Fig. 4.5. Historical data is used as input to the KDD process that outputs patterns, subjected to interestingness measures (i.e., defined quality) [175]. An interestingness measure can either be objective (i.e., based on statistical strength/properties) or subjective (i.e., based on the user's beliefs/expectations) [175]. A pattern represents a subset of process data in an abstract way, thus requires the evaluation of a domain expert (e.g., plant manager) to confirm a pattern and identify knowledge. The steps of the KDD process are depicted in Fig. 4.5 and are elaborated in the following paragraphs. The understanding of the considered manufacturing domain (e.g., aluminum sand casting) is a prerequisite for a successful employment of the KDD process. The process analysis and (re-) design in Sect. 4.2.1 as well as the enterprise data model and DFDs in Sect. 4.2.2 describe methodologies that foster the comprehension of the manufacturing domain in concern. In the context of this research work, the aim of the KDD process is to identify knowledge that can be used for real-time monitoring and control of manufacturing processes. Manufacturing activities can be assigned to production, maintenance, quality, or inventory, as defined by IEC 62264 (cf. Fig. 3.9 in Sect. 3.3.2.3). Henceforth, these activities are the basis to define the goals of the KDD process. The selection of target data is done depending upon these KDD goals. Hence, target data is a subset of process data that is searched for patterns.

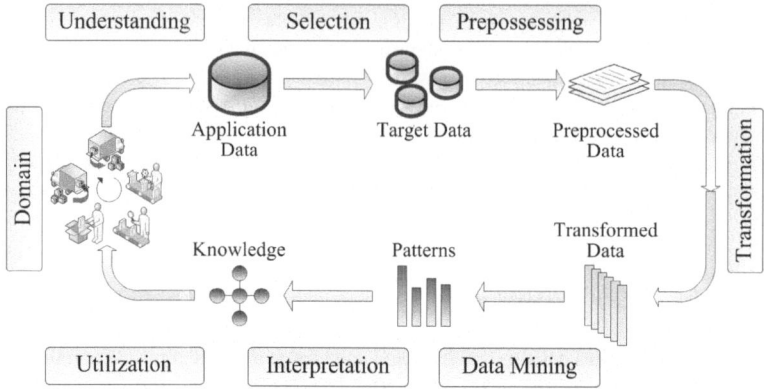

Fig. 4.5 Knowledge Discovery in Databases (KDD) process (adapted from [71]).

Inaccuracies in historical process data can occur for several reasons, thus might adulterate the derived knowledge. Process data can be collected using various techniques, for instance: (i) workers' feedback inputted *manually* into terminals; and (ii) machine and product data gathered *automatically* from automation devices[6]. The collected process data can possess noise, inaccuracies, and missing values, which all hamper the proper detection of patterns [203].

The reasons for these nuisances are numerous: (i) limited accuracy of measurement instruments; (ii) typos in workers' manual feedback; and (iii) logical errors in the PLCs. Understanding the considered manufacturing domain along with knowledge about interdependencies between IT-systems and resources are crucial to identify suitable statistical methods to (i) remove noise; (ii) define a strategy to fill missing values; and (iii) delete duplicate data. Statistical methods might also disclose shortcomings of the implemented data collection methodologies. For instance, the inspection of the quality of the determined parameters can be checked using histogram-based methods [50]. Subsequently, appropriate measures can be initiated to mitigate quality shortcomings. In summary, the selected process data is cleaned and pre-processed before it is used for further processing.

Often, only a subset of the pre-processed data is necessary to retrieve patterns (i.e., new knowledge) with regard to the KDD goals [232]. As not every parameter is useful for identification of new knowledge, irrelevant parameters can be removed employing filter and wrapper approaches [84]. Further, target data can be transformed into a more general or acceptable format. Also, comprehension of the manufacturing processes, operations, and constraints, which are reflected in the enterprise data model and DFDs, fosters this transformation process.

Data mining is a sub-process in the KDD process, and is based on established techniques, like machine learning, pattern recognition, statistics, artificial intelli-

---

[6] A module for automatic collection of process data is elaborated in Sect. 4.3.1.

gence, knowledge acquisition, data visualization, and high performance computing [109]. The data mining sub-process is composed of three steps: (i) choice of a data mining method; (ii) selection of an appropriate data mining algorithm; and (iii) execution of data mining algorithm on transformed data for identification of patterns.

Data mining methods determine the type of knowledge to be mined. Knowledge types are, for instance, concept description, classification, association, clustering and prediction [44]. Hence, they have to be selected with respect to the intended KDD goals. Classification and regression methods support decision-making processes, thus are promising candidates to enhance real-time monitoring and control of manufacturing processes.

Classification is employed to group new instances of data into predefined classes [71, 203]. Classification is composed of two steps [44]: (i) derivation of a *classification model* based on the analysis of training sets composed of process data tuples; and (ii) *assignment* of new process data instances to certain classes according to the designed classification model. In contrast, *regression analysis* is used to reveal the quantifiable relationship of a dependent variable with other independent variables. The outcome is a regression function that can be employed to predict the real value of the dependent variable assuming certain values of independent variables.

Based upon the previously selected data mining methods (e.g., regression analysis), specific data mining algorithms have to be chosen to implement these data mining methods. A plethora of data mining algorithms exists to determine rules, namely decision trees, decision rules, inductive logic programming and rough set methods [153]. Overall, the data mining methods and algorithms are used to determine patterns that represent new knowledge.

The discovered patterns and relationships can be large, thus demand an evaluation of these patterns. The interestingness of a certain pattern can be measured, thus the quality of a pattern can be assessed in an objective way [84]. In addition, domain experts can be interviewed to evaluate the discovered patterns. These interviews are indispensable if there is not sufficient process data available. In some instances, it is "sufficient to perform only structured interviews with domain experts to identify critical control-related process parameters" [97, 403].

Usually, the aforementioned KDD process gets repeated regularly to obtain more suitable knowledge. Over time, the knowledge base can be enlarged and enhanced by (re-) performing parts of the KDD process. Also, a repetition of the KDD process can be exploited to confirm or discard previously generated rules.

4.2.4 Knowledge-Based Monitoring and Control of Manufacturing Processes

As mentioned in the previous subsection, process data can be collected from the shop floor and managed as historical data in a relational database. A KDD process along with structured interviews can be employed to identify control-related parameters and derive new knowledge for monitoring and control of manufacturing processes. Further, this knowledge can be employed in CIPs. Although this approach for monitoring and control of manufacturing processes is expedient, it is purely *offline* and cannot provide immediate (re-) actions on manufacturing process disturbances.

Obviously, a repeated querying of the relational database might offer a solution for *online*, i.e., near real-time, monitoring and control of manufacturing processes. However, repeated queries are time consuming and require a lot of computational resources [38]. In the context of this research work, a more advantageous approach based on *in-memory processing* of process data is favored. This real-time monitoring and control is part of a comprehensive IT framework that realizes a vertical integration of different enterprise control levels. Different perspectives, especially, concepts of management, engineering, and the computer science community are incorporated. Further, reasonable ideas and models are adopted from research on intelligent manufacturing control (cf. Sect. 4.1.2).

Following, the working principle of the developed control approach is summarized[7]. This working principle is based on three consecutive steps: (i) in-memory management of *tracking objects* as representatives of process entities; (ii) real-time *analysis of tracking object streams* employing CEP for detection of (critical) process situations; and (iii) deduction of appropriate *(re-) actions* to improve the manufacturing processes.

A tracking object can be interpreted as a *virtual image* of an *individual* process entity. This process entity can represent a semi-finished product, a production order, and so forth (cf. the entities mentioned in the context of holons in Sect. 4.1.2.4). Each tracking object is constructed according to the definition of an event object as explained in Sect. 2.2.2. Henceforth, a tracking object can also be interpreted as a *complex event* that encompasses event form, significance and relativity (cf. [161]). Noteworthy, a tracking object is composed of tracking object items, i.e., event object attributes, which are gathered and *integrated* from different enterprise control levels. Technically, tracking objects are totally managed in the *main memory* of the control system.

During execution of manufacturing processes several process activities are performed that influence individual process entities. Consequently, tracking objects are constantly updated using collected process data, and thereby, provide up-to-date status information about an individual process entity. Further, the tracking objects are streamed into a CEP engine that interprets the tracking objects as (complex) events. These streams are analyzed employing rules that represent knowledge identified during the KDD process (cf. Sect. 4.2.3). Finally, appropriate (re-) actions are initiated after the detection of a critical process situation: (i) sending of an alarm message to an operator/manager; (ii) presentation of advice to an operator/manager; or (iii) dispatching a control command to directly influence a shop floor resource/IT-system.

---

[7] A more detailed description of the IT framework, its software components and their interdependencies is outlined in the upcoming sections.

4.3 Event-Driven Framework for Real-Time Monitoring and Control

The elaborated process model is a mandatory precondition for the implementation of the IT framework that is presented in subsequent paragraphs. The framework has primarily been designed for the realization of the RTE in manufacturing. As a prerequisite for the RTE, the framework follows the implementation of a vertically integrated enterprise that bridges the temporal and semantic integration gap (cf. Sects. 2.1.1 and 2.1.2). Based on this integration of several enterprise control levels, it realizes a real-time monitoring and control of the value creation processes (i.e., manufacturing processes). The addressed application area has been characterized in Sect. 3.1.

4.3.1 Outline of Architectural Components

The IT framework has been developed in a real industrial environment, thus differs from most laboratorial research, as delineated in Sect. 4.1.2. Further, the framework has been presented in journals and conferences, and has been discussed with experts from management, engineering, and the computer science community. The following outline of the architectural components of the framework expands on [93,95,96].

An overview of the developed IT framework is depicted in Fig. 4.6 (cf. [96]). At the logical process layer, several instances of (running) manufacturing processes are depicted. These process instances are composed of a sequence of process steps. A process model for certain manufacturing process types was created during analysis and (re-) design of these processes (cf. Sect. 4.2.1). According to the description of the application area (cf. Sect. 3.1.1), the process steps, i.e., processing operations, assembly operations, and logistic operations, are executed primarily by machines and automation devices. The IT framework is designed on top of this process layer. The IT framework is composed of software components that communicate following the event-based interaction model (cf. Sect. 2.2.5). The main components of the IT framework are: (i) *data collection engine* for the integration of machines and automation devices located at the manufacturing level; (ii) *data aggregation engine* that relates transactional and real-time process data from different control levels; (iii) *online tracking* of process entities as part of the data aggregation engine; (iv) real-time monitoring of manufacturing process instances facilitating CEP, and subsequently, dispatching of control commands to achieve the objectives set at the enterprise control level; and (v) visualization of process data and tracking information in real-time as well as a provision of forward and backward traceability of manufacturing processes.

The aforementioned components are seen as compulsory for the realization of multiple closed loop controls and the realization of knowledge feedbacks within and across enterprise control levels [146, 222]. Hence, the IT framework represents

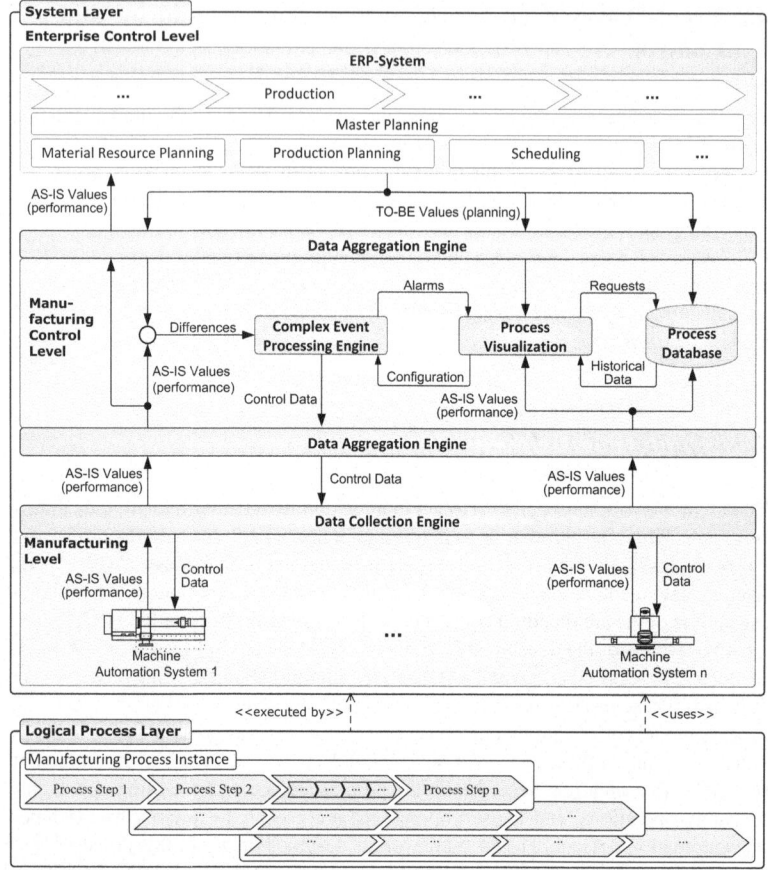

Fig. 4.6 Overview of event-driven IT framework for establishment of RTE in manufacturing (adapted from [96]).

a step toward the RTE [99]. The above software components are elaborated in subsequent paragraphs. However, the focus is set on online tracking and the real-time monitoring and control of manufacturing processes employing CEP.

4.3.2　Real-Time Acquisition of Process Data from Manufacturing Resources

The previously described manufacturing processes are supported with numerous automation devices, such as special purpose machines, conveyor belts, robots, and manipulators. These automation devices are located at the manufacturing level and are connected using various industrial communication standards, like PROFINET, PROFIBUS, and proprietary protocols, as illustrated in Fig. 4.7. These devices are arranged in a certain configuration as defined in the plant layout. Advances

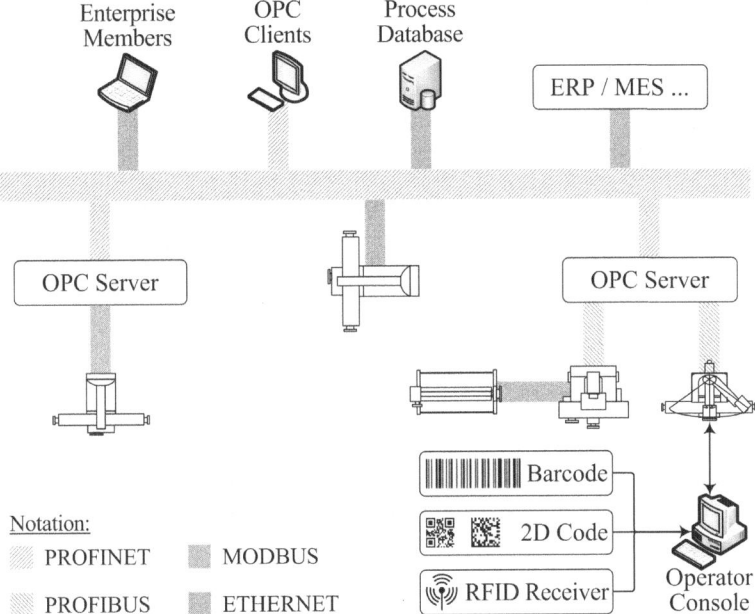

Fig. 4.7 Automation devices and manufacturing management systems connected employing different (industrial) networks.

in industrial IT have made it possible to integrate PLCs into most of the devices, with/without dedicated terminals, which assist in monitoring and control of manufacturing processes. In the case of manually operated machines, special terminals/-consoles need to be made available to the operators of these machines, which assist the bi-directional communication with other levels of the control hierarchy. Further, the automation devices are complemented with the manufacturing support systems, such as MES.

The automation devices generate real-time process data while executing the manufacturing processes. The PLC of an automation device is implemented by the mechanical engineer. The capability to acquire process data is crucial for real-time monitoring and control of manufacturing processes. Therefore, the data collection engine is the basis for additional software components. The architecture of the data collection engine[8] is depicted in Fig. 4.8.

A plethora of standard and proprietary protocols exist to communicate with devices at the shop floor level, as illustrated in Fig. 4.7. For instance, OPC servers have been implemented to gather process data from PLCs. OPC has its origins in the process visualization community that searched for a standard communication protocol to access process data. As most visualization tools were based on Windows® platforms, OPC was developed on top of the Microsoft™ DCOM protocol (cf. also the historical background of EP in Sect. 2.2.4). This OPC version is known as OPC Data Access (DA).

The latest development from the OPC foundation has led to the OPC Unified Architecture (UA) that overcomes some limitations and shortcomings of OPC DA (cf. [80]). Firstly, OPC UA is based on SOA, thus fosters interoperability among devices and control levels [80]. Secondly, a lean binary TCP-based binary protocol has been developed to fulfill strong performance requirements [80]. In addition to OPC, standard TCP/IP protocols (e.g., Modbus) and proprietary protocols for exchanging data (e.g., writing files to a network drive or sending data using raw TCP/IP packets) are also common in a scenario in which different automation devices are running.

Because of the diversity of machine protocols, the data collection engine has been developed in a modular way, as depicted in Fig. 4.8. Various communication protocols are available in a protocol manager that can be extended with new protocols. Each protocol is implemented by a driver that is used in the data acquisition manager. The data acquisition manager is configured by a configuration manager. Process data items are members of one or more process data groups. In most instances, a process data group describes a certain activity in the manufacturing process. Trigger conditions are defined for each of these process data groups. If a trigger condition is completely fulfilled, the process data items of the corresponding process data group are gathered[9]. To achieve high data throughput and low latencies, the data acquisition manager creates a separate thread for each process data group. The process data is acquired using a certain driver, pre-processed (e.g., cleaned or filtered), and forwarded to the dispatching manager on activation of a trigger. Different trigger modes for a process data group have been identified. First, all process data items are read continually using a polling interval of a predefined length (e.g., 1000 ms). This communication pattern is termed by computer science community as *polling*. Next, individual process data items of a process data group are collected when there is a change of the corresponding process data value. This

---

[8] The initial version of the data collection engine was presented by [99].

[9] Although standard protocols are employed, the internals of a PLC are dictated by the corresponding mechanical engineer. Hence, the connection to a machine or automation device necessitates an engineering process to specify the communication interface in detail.

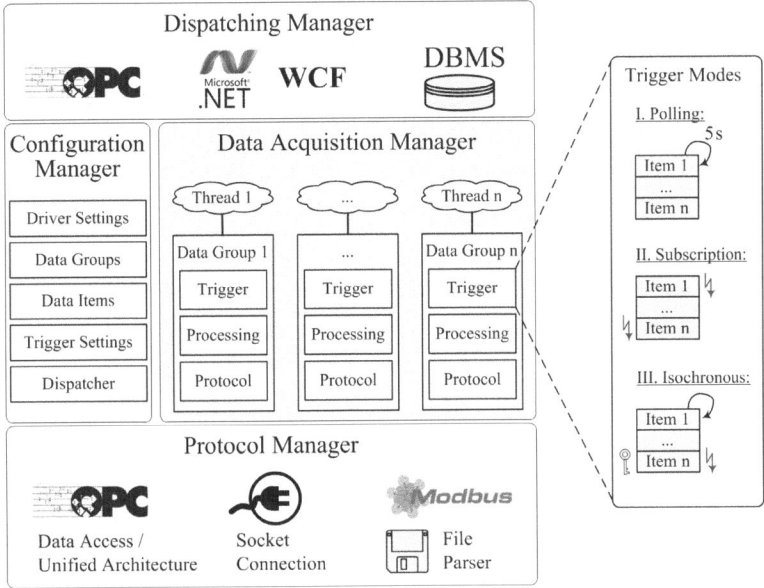

Fig. 4.8 Illustration of data collection engine for integration of shop floor devices.

approach is based on *publish-subscribe* mechanism (cf. Sect. 2.2.5) that is preferred for real-time control activities.

Finally, an *isochronous* communication pattern can be implemented (cf. [138]). In this case, a process data group is composed of a *primary key* and process data items, which are summarized as *payload*. The publish-subscribe mechanism is employed for requesting the primary key of the process data group. Upon a change of the primary key, the associated payload is read from the device using the request-reply pattern. The isochronous communication pattern is not as rigid as polling, but not as lenient as the publish-subscribe mechanism. The appropriate trigger mode is configured based on the circumstances of the device and its programming. The trigger mode is defined for each process data group within the configuration manager.

The dispatching manager receives the collected process data and can send it to other (software) components. Several communication protocols, like Microsoft[TM] Windows Communication Foundation (WCF) and OPC, can be incorporated to interoperate with different receivers. Noteworthy, event-based communication can be realized with the dispatching manager, thus process data is transferred immediately to registered software components.

4.3.3 Aggregation of Process Data for Forward and Backward Traceability

The data aggregation engine is a software component that receives process data from the data collection engine and integrates this shop floor data with transactional data from overlying enterprise control levels. Thus, the collected process data can be interpreted and assessed according to its technical and non-technical context. As de-

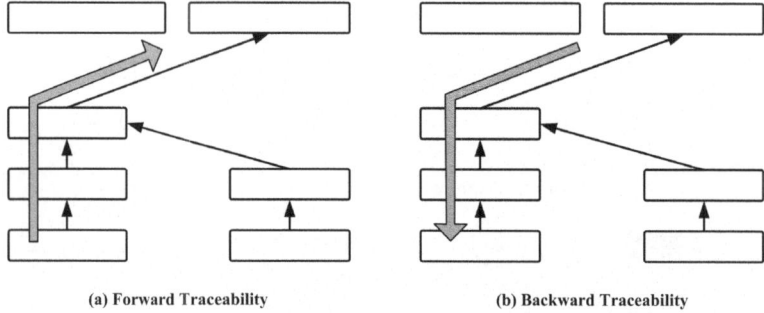

(a) Forward Traceability   (b) Backward Traceability

Fig. 4.9 Illustration of forward and backward traceability (adapted from [143]).

scribed in the process model (cf. Sect. 4.2), integrated process data that is managed in a relational database is helpful to investigate manufacturing processes' efficiency. The recorded process data is also suitable to identify new knowledge for improvement of the manufacturing processes. This knowledge can be employed, especially, to generate rules for enhancement of real-time monitoring and control of manufacturing processes. In the context of MES, the stored process data is employed to realize the *traceability* functionality (cf. Sects. 3.3.2.1, 3.3.2.2, and 3.3.2.3).

Traceability is defined as an "ability to preserve the identity of the product and its origins or more vividly as a 'possibility to trace the history and the usage of a product and to locate it by using documented identification" [250, 8]. Similarly, *tracing* refers to "storing and retaining the manufacturing and distribution history of products and components" [139, 546]. VDI 5600 describes traceability as composed of the following sub-processes: (i) documentation of production flows; (ii) acquisition of process data for complaint verification; (iii) acquisition of product data; (iv) analysis of product and process data for different objectives; and (v) archiving of product and process data [255]. Overall, tracing is an activity that is performed retrospectively.

Traceability can be accomplished on various enterprise entity types, like product, quality, material lot, production order, production plan, machine, operation, and personnel [145]. Further, traceability can be split into forward and backward traceability, as depicted in Fig. 4.9. *Forward traceability* answers the question of where a particular process entity has been used (i.e., material implosion). Contrary, *backward traceability* identifies the process entities, like sub-products and materials, which have been consumed by a particular enterprise entity (e.g., individual product) in consideration (i.e., material explosion).

The acquired process data has to be mapped onto the aforementioned process entities that, in turn, have to be stored in a relational database. An *enterprise reference model*[10] describes the relationships between the process entities. Hence, it results in an enterprise data model that is the basis for the database schema of the relational database. An example of an enterprise reference model is depicted in Fig. 4.10. Usually, the enterprise reference model has to be adapted according to the enterprise in consideration. Traceability is crucial to analyzing manufacturing pro-

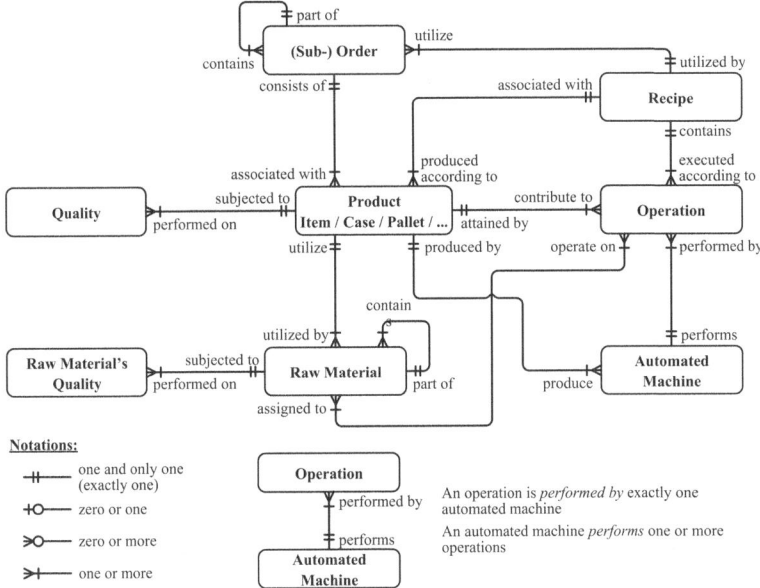

Fig. 4.10 Example of an enterprise reference model for aggregation of process data, depicting relationships among process entities (modelled using crow's foot notation of entity-relationship diagram).

cess disturbances (e.g., identification of causes for loss of production). However, traceability can be achieved at different granularity/resolution levels, as depicted in Fig. 4.11. Basically, traceability can be established at two levels–*unit/item level* and

---

[10] In the realm of traceability research the synonym *reference data model* is used (e.g., [56]). Also, the term *enterprise data model* can be used, as in Sect. 4.2.2. The terms simply provide different views on the data/process entities and their relationships. Henceforth, an enterprise can have multiple views on its data/process entities and their relationships, and be drawn using different notations.

lot/batch level [145]. The traceability resolution often depends on the structure and shape of (intermediate) goods/materials (cf. Fig. 3.2 in Sect. 3.1.1). Item and lot resolution are both used in complex manufacturing processes that are characterized by the coexistence of multiple types of manufacturing (e.g., batch and discrete manufacturing) [189]. In some instances, it can be necessary to employ a *date-code* to identify and trace a certain process entity. Further, the lot resolution can be subdivided dependent on the container (e.g., pallet, case). The item resolution often

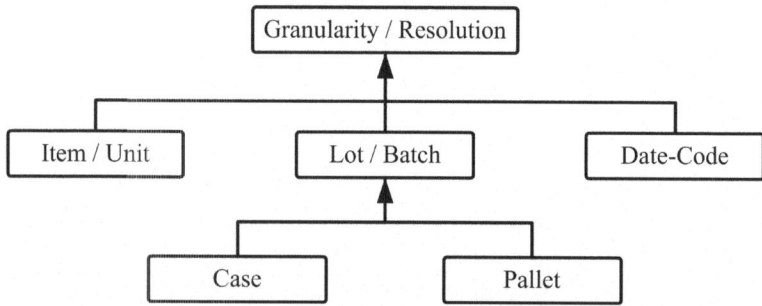

Fig. 4.11 Classification of granularity/resolution levels for traceability.

demands relatively high storage capacities as more individual process entities are recorded. In addition, the recording frequency is higher compared to the lot resolution. The lot resolution, however, can result in complicated tracing operations as lots can be nested or merged.

An overview of the data aggregation engine to establish the aforementioned forward and backward traceability is given in Fig. 4.12. In addition, the data aggregation engine is the basis for real-time tracking of process entities, thus implements a tracking object manager. However, the focus in the current section is on the storage and management of integrated process data. Although the concept of tracking of process entities is outlined, it is described in detail in the upcoming section. The data aggregation engine subscribes to the data collection engine (cf. Sect. 4.3.2) that, in turn, continually provides process data. The process data from the manufacturing level denotes AS-IS values of the manufacturing processes and corresponding process entities. The AS-IS values are integrated by the data aggregation engine with related TO-BE values from the enterprise control level (e.g., ERP system). Further, the integrated process data is stored in a relational database that is also called a process database. The details of the above functionality of the data aggregation engine are elaborated.

On start, both the data collection engine and the data aggregation engine are initialized from an XML-based *configuration file*, as depicted in Fig. 4.12. This file can be generated with an editor. The configuration file defines information that is required by the data aggregation engine during run time. The information describes

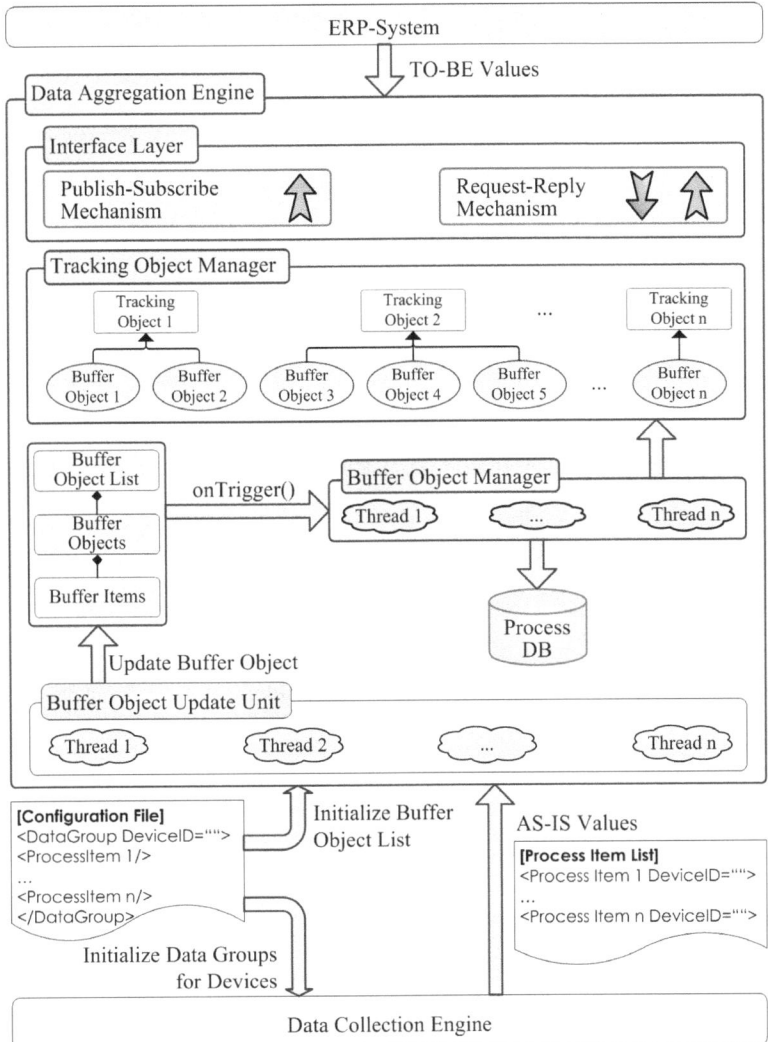

Fig. 4.12 Overview of aggregation engine as basis for (i) forward and backward traceability; (ii) real-time tracking of process entities; and (iii) control of manufacturing processes (adapted from [138]).

manufacturing resources, data blocks of PLCs, buffer objects, process data items, buffer items, database settings, and so forth.

A manufacturing resource (e.g., machine or automation device) is often equipped with a PLC that has a communication interface to the data collection engine. Each resource is identified by a unique device identification tag (UDIT). A resource has one or more process data groups, i.e., data blocks within a PLC, which are linked to the UDIT. In most instances, a process data group denotes to a possible operation performed by a certain resource. A data group contains process item definitions, trigger conditions, and parameterization of machine communication. Each process item is described by a name, address information of the PLC, and a data type.

The process items are members of one or more buffer objects that are managed by the *buffer object manager*. Usually, a buffer object represents a certain data group in an automation device (e.g., PLC). A buffer object is composed of several buffer items that are similar to process items. In contrast to the process items, buffer items encompass additional information, like readable names (e.g., "Temperature Sensor 1"), column name for mapping onto a database table (e.g., TEMP_1), and a physical unit (e.g., ° C). The configuration file explicitly specifies the relation between buffer items and a corresponding buffer object. It is defined when (i.e., trigger condition) and where (i.e., database table name) a buffer object has to be stored in the relational process database.

The buffer objects are initialized on start of the data aggregation engine with default values for the buffer items (e.g., empty string). The buffer objects are kept in the main memory of the engine and are managed in a buffer object list. The data collection engine pushes process item lists, which are composed of process items, to the data aggregation engine. A *buffer object update unit* processes different process item lists concurrently in separate threads to ensure high throughput. A process item is mapped onto a corresponding buffer item[11]. The interaction of buffer items and a buffer object is shown in Fig. 4.13. A buffer object is *created*[12] with the first incoming buffer item that belongs to this buffer object. Afterward, the buffer object is updated with every new buffer item. Thus, only one region of main memory has to be allocated for a certain buffer object, as a recurring process operation just initiates an update of the existing data structure (i.e., buffer object). Buffer items arrive at different times and frequencies, thus result in an asymmetric update process. Further, the data aggregation engine implements interfaces to IT systems located at the enterprise control level (e.g., ERP system), and thus, can access transactional data

---

[11] In most cases, a process item is mapped onto exactly one buffer item. However, it is also possible to use a process item for more than one buffer item.

[12] From a programming perspective, the buffer object is already *created* as an object in main memory during the aforementioned initialization phase. Therefore, the term "created" is used to indicate the very first appearance of the process operation that is signified by the buffer object.

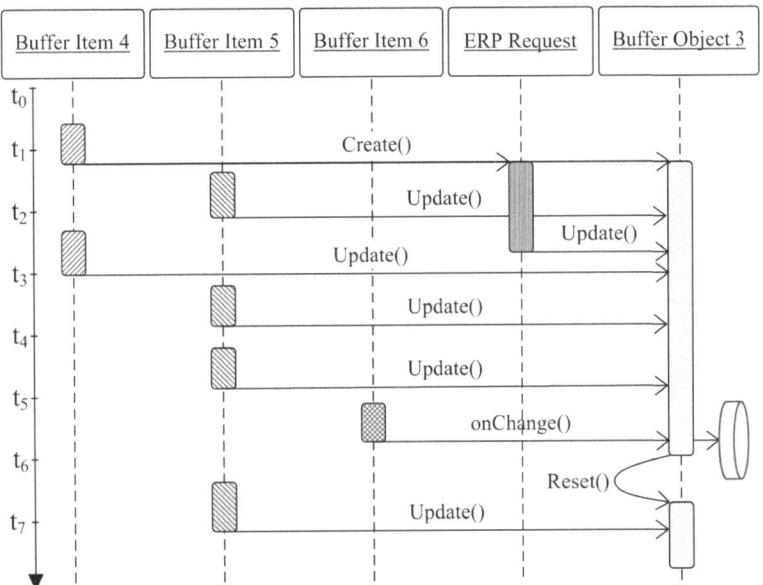

Fig. 4.13 UML sequence diagram depicting the interaction between buffer items and buffer object [93].

(i.e., TO-BE values) from these IT systems[13]. These transactional data can also be mapped onto buffer items of a buffer object.

After the fulfillment of a trigger condition, a buffer object has to be committed in the process database. For that purpose, a snapshot of the buffer object is taken by the buffer object manager. Further, this snapshot is processed in a separate thread and an *upsert statement* is executed on the relational process database. An illustration of how a buffer object with associated buffer items is mapped onto a table in the relational process database is given in Fig. 4.14. A buffer object contains information about the database table and primary key(s). In addition, the buffer object encompasses properties that specify trigger conditions–when to take a snapshot of the buffer object inclusive of its buffer items. The update of a buffer item is performed by the invocation of the method `SetBufferItem(key, value, timestamp)`. A buffer item is assigned to a buffer object, and further, includes a column name and process data value. The information about the database table, primary keys, column

---

[13] Even though the presented research work assumes that the ERP system is the central enterprise application at the enterprise control level, product-related data provided by a PDM system might also be integrated with process data from the shop floor. In addition, the acquired process data (e.g., documentation about how an individual product has been manufactured) is an essential part of a product's lifecycle [240]. Similarly, Thiel suggests incorporating PLM as part of an integrated production management system including ERP, PLM/PDM, and MES [248].

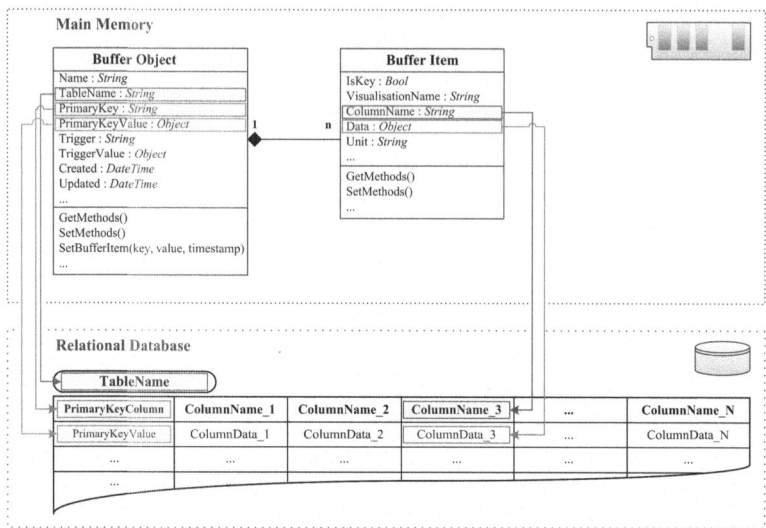

Fig. 4.14 Simplified illustration of working principle for mapping of a buffer object with buffer items to a database table in the relational process database.

names and process data values are used to either *create* a new row for this database table or *update* a row of this database table.

An example of showing the mapping of three different buffer objects onto a database table is depicted in Fig. 4.15. The database table Produced Products stores information about actually executed manufacturing operations for individual products that have been produced on a certain production line. Each product is identifiable with a unique Product ID. The following sequence of manufacturing operations is executed: (i) cutting of material from an iron cube; (ii) assembly of a screw; and (iii) quality inspection of assembled product. After completion of the cutting operation, the corresponding buffer object is mapped onto the table Produced Products and a new row for product 4711 is created. Thereafter, the assembly and inspection operations just update the row for product 4711. Consequently, the residual data structure is *flat*, thus accelerates the access of individual process entities (e.g., product with product ID). Further, the clarity of the data structure makes it more readable and understandable. In essence, the flat organization has advantages while analyzing the process data and tracing process entities (cf. discussion about column- and row-oriented databases in Sect. 2.2.4). In contrast, available *data historians* focus on storing process data according to time series. Because of the high frequency of sensor data, they are optimized to record data in (near) real-time. Moreover, data historians provide data compression mechanisms to handle huge data volumes. However, they are not optimized for managing process entities, which are described by many interrelated properties.

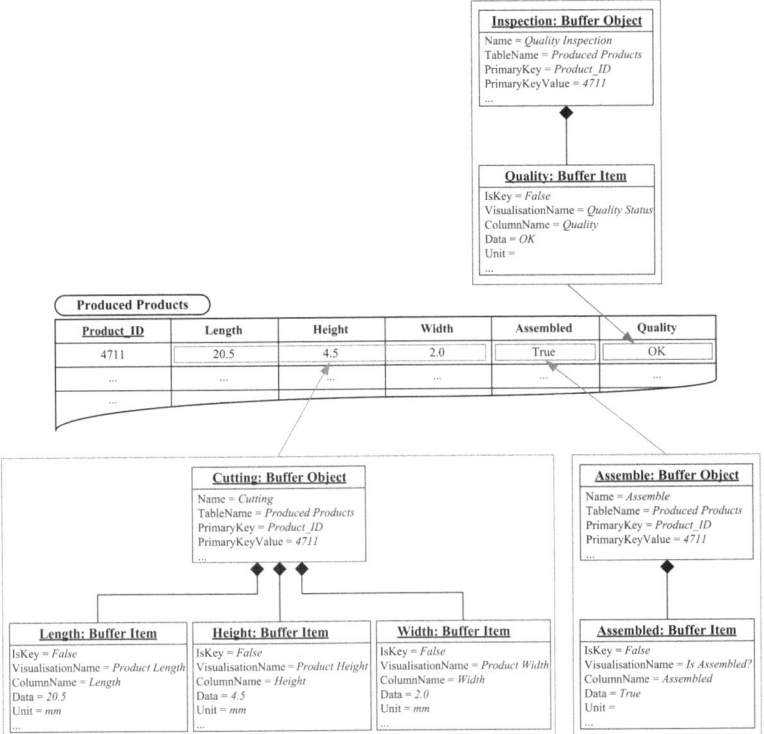

Fig. 4.15 Example showing the mapping of three different buffer objects onto a database table.

As mentioned earlier, the storage of a buffer object is provoked by the fulfillment of certain trigger conditions. The trigger conditions depend on the engineering and implementation of the automation device's PLC, which defines conditions indicating when process data in a data block is stable for getting read. Three different trigger conditions have been designed[14]: `onChange()`, `always()`, and `onCondition()`.

The `onChange()` trigger condition is fulfilled, if a predefined buffer item's value has been changed. The `always()` trigger condition is employed to continually commit a buffer object to the relational process database on the arrival of a buffer item. This trigger condition is often used by data historians for storing time series. The `onCondition()` trigger condition is an extended version of the `onChange()` trigger condition. In this case, a buffer object is only committed, if a certain buffer item has changed to a predefined value. In summary, buffer items are

---

[14] This list is not complete, thus can be extended by other trigger conditions.

updated continually, while buffer objects are stored only on satisfaction of trigger conditions.

Finally, the data aggregation engine manages connections to various software components of the IT framework, as depicted in Fig. 4.12. It implements interfaces to visualization clients that, in turn, provide reliable and convenient ways to access the stored process data. The management of buffer objects and related buffer objects as well as the committing of buffer objects to the relational process database have been conceptualized as being performed in separate threads. Therefore, it is necessary to implement the data aggregation engine functionalities in a thread-safe manner (e.g., mutual exclusion).

4.3.4 Online Tracking of Enterprise Process Entities

The aforementioned management of process data and its integration with transactional data is the base for (i) forward and backward tracing of process entities; and (ii) identification of knowledge concerning monitoring and control of manufacturing processes. However, both tasks are performed *retrospectively*, thus are unsuitable for monitoring and control in (near) real-time. Therefore, in the following paragraphs, the concept of *online* tracking of process entities is introduced. Online tracking is a means to monitor process entities *during* execution of manufacturing processes. As such, it is a starting point to analyze manufacturing processes and deduce appropriate (re-) actions.

Tracking is an "act of observing, in most cases, the spatial motion of an entity" [143, 4494]. Similarly, tracking can be defined as "gathering and management of information related to the current location of products or delivery items" [139, 546]. A broader definition is provided by IEC 62264-3 in which tracking signifies the "activity of recording attributes of resources and products through all steps of instantiation, use, changes and disposition" [127, 10]. Similar to tracing, tracking can be achieved with different granularity levels, i.e., item/unit and batch/lot level [145].

As elaborated in the previous section, buffer objects are committed to the relational process database on the fulfillment of predefined trigger conditions. Concurrently, these buffer objects are forwarded to a *tracking object manager*. The tracking object manager is in charge of managing the life cycle of tracking objects (i.e., creation, update, and deletion). A tracking object can be interpreted as the virtual image of an (individual) physical or non-physical process entity (e.g., individual product, production order).

The structure of a tracking object is depicted in Fig. 4.16. A tracking object contains data members, which can be separated into three categories: (i) data required for the management of the tracking object; (ii) data employed to derive the status of the corresponding process entity (i.e., tracking object items); and (iii) references to other tracking objects. The first category encompasses data, like a primary key/unique identification tag, timestamps, lifetime, and so forth. Usually, either the primary key alone or the combination of tracking object type and primary key is a

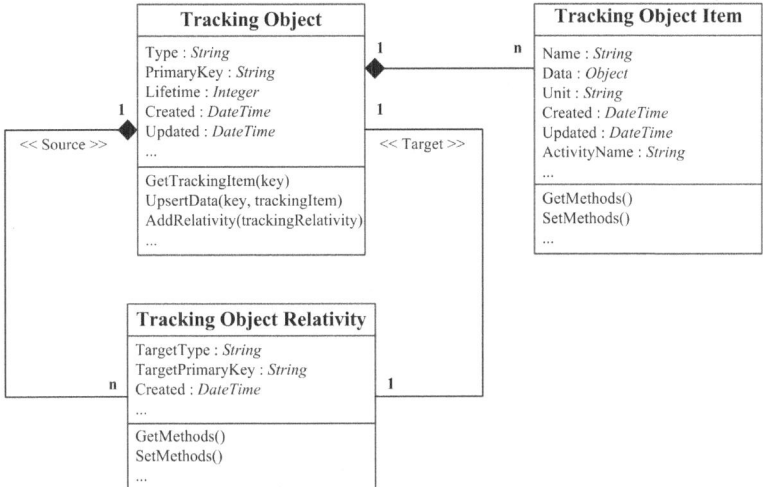

Fig. 4.16 Tracking object with associated tracking items and relativities to other tracking objects, modelled as UML class diagram.

unique identifier for a tracking object. In the second category, associated tracking object items describe the status of the process entity. Each tracking object item is associated with a process activity (e.g., manufacturing operation). Hence, a process activity is described by a number of tracking object items that, in turn, belong to a certain tracking object. Finally, a tracking object can have multiple references to other tracking objects. The relationships between tracking objects are modeled with tracking object relativities.

The tracking object is composed of condensed status information about a process entity. This information can be related to the process entity's location, operations performed on this process entity, quality inspections, and so forth. As mentioned earlier, a buffer object describes a process activity (e.g., manufacturing operation) with a set of buffer items. After receipt of a buffer object, the tracking object manager maps a subset of buffer items onto tracking object items of relevant tracking objects. The restriction of this mapping to a subset of buffer items is motivated by the fact that (i) tracking objects are kept in main memory, which is a scarce resource; and (ii) only items are relevant that are useful for real-time monitoring and control of manufacturing processes. In summary, the relationships among process item, buffer item, buffer object, tracking object item and tracking object are illustrated in Fig. 4.17. The KDD process of the process model, mentioned in Sect. 4.2.3, can be employed to identify control-relevant parameters, and subsequently, is the basis for ascertainment of tracking object items. A definition of tracking objects is provided in a XML configuration file that explicitly describes the relations among buffer items, buffer objects, tracking object items and tracking objects. The refer-

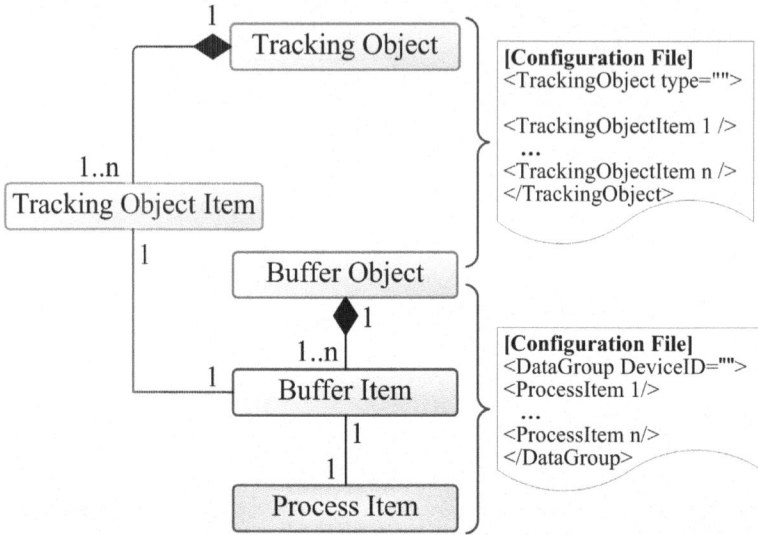

Fig. 4.17 Relationships among process item, buffer item, buffer object, tracking object item and tracking object (adapted from [93]).

ences between tracking objects are specified by tracking object types and tracking object primary keys (cf. Fig. 4.16).

An example of a tracking object "Product A", which includes subsets of buffer objects and references to other tracking objects, is depicted in Fig. 4.18. Manufacturing operations are performed at different times, denoted as $@Tx$, $x \in \mathbb{N}$, and corresponding buffer objects are created or updated. Subsequently, a subset of buffer items is extracted from these buffer objects and mapped onto tracking object "Product A". Further, references between other tracking objects are set during manufacturing process execution. Along the execution of manufacturing processes, a tracking object undergoes different phases of life cycle - creation, modification and destruction. The situations, when a tracking object has to be created and modified are explicitly defined in the aforementioned XML configuration file. Tracking objects are created or updated on receipt of corresponding buffer objects. An area of the system's main memory is allocated on creation of a tracking object. Further, links are established between tracking objects with execution of certain manufacturing operations and concurrent creation/update of buffer objects.

As tracking objects are completely kept in main memory, it is critical to define the termination condition, especially for transient process entities, like products and orders. A tracking object is purged by removing the reference to this object and freeing the allocated memory for new tracking objects. Releasing the allocated

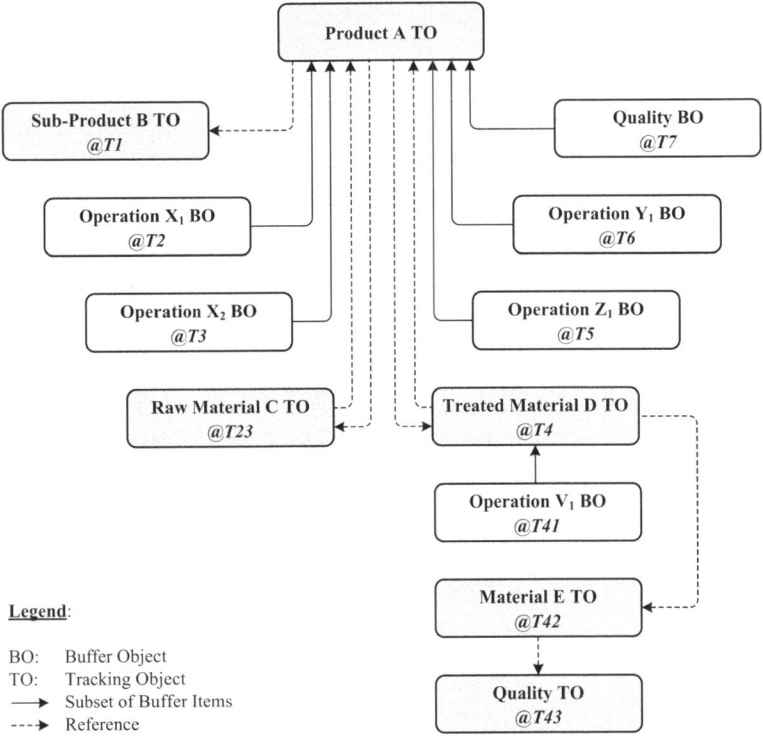

Fig. 4.18 Illustration of tracking object of type "Product A", assignment of buffer items signifying process operations, and relationships between tracking objects of different types.

memory area enhances the performance of the system in terms of memory footprint and execution speed.

The conditions for termination of a tracking object can be defined in numerous ways. The production schedule (cf. Sect. 3.3) contains order information with regard to quantity, BOM, material routing, resources, and so forth [238]. The last step of the material routing can be defined as a termination condition. Also, a workflow step that is associated with an order can be used as a condition for destruction of the corresponding order tracking object. Finally, a maximum expected lifetime is defined for each tracking object (cf. Fig. 4.16). A special garbage collector manages the destruction of expired tracking objects employing the specified termination conditions.

Timestamps are set or updated on creation/update of a tracking object. These timestamps are important for further processing of the tracking objects, for instance, within a CEP engine. In some cases, it is essential to distinguish the instant in time when an activity has been performed, i.e., its application time, from the instant of time when it is detected/received by another system, i.e., its system time (see also the discussion on detection and occurrence times in Sect. 2.2.3). The capability to capture and handle application timestamps becomes critical in scenarios in which *hard* real-time requirements (cf. Sect. 3.4) have to be fulfilled.

Tracking objects integrate transactional data from the enterprise control level with process data from the manufacturing level. Therefore, tracking objects are used to generate reports in (near) real-time. These reports are tailored for different enterprise members, like supervisors, plant managers, and workers. Each enterprise member requires pertinent perspectives and visualizations of the tracking objects. Privileges to access tracking objects can be managed, for instance, on a lightweight directory access protocol (LDAP) server.

There are several possibilities of how one can navigate through tracking objects and tracking object items, as summarized in Fig. 4.19. Navigation between tracking objects is realized using tracking object relativities (tracking object → tracking object). Also, tracking object items can be resolved for an individual tracking object (tracking object → tracking object item). Further, tracking objects can be identified that possess a particular tracking object item (tracking object item → tracking object). Finally, navigation between tracking object items can be realized (tracking object item → tracking object item). Overall, tracking is a tool for enterprise

Input \ Output	Tracking Object	Tracking Object Item
Tracking Object	Tracking Object → Tracking Object	Tracking Object → Tracking Object Item
Tracking Object Item	Tracking Object Item → Tracking Object	Tracking Object Item → Tracking Object Item

Fig. 4.19 Possibilities to navigate through tracking objects and tracking object items (adapted from [250]).

members to track process entities in (near) real-time. Thereby, tracking objects provide an integrated view of process entities encompassing non-financial/technical and financial process data. This up-to-date information can be analyzed by enterprise members to proactively control the manufacturing processes. However, this is a manual act that is not performed in real-time. Therefore, the tracking objects are concurrently forwarded to a CEP engine for real-time monitoring and control of manufacturing processes.

4.3.5 Detection of (Critical) Process Situations using CEP

A comparison of a tracking object with an event object is summarized in Fig. 4.20. Overall, a tracking object is designed like an event object. However, while an event object encompasses attributes of a single process activity, a tracking object integrates attributes of several process activities into a single data structure. Hence, a tracking object can be interpreted as an aggregated or complex event (cf. Sect. 2.2.3), which can be automatically analyzed by a CEP engine.

Object Element	Event Object	Tracking Object
Significance	Signifies an activity	Signifies a process entity that is updated by various activities
Attributes	Attributes / data components to describe the activity	Attributes to describe the process entity; including attributes of performed activities
Relativities	Relativities to other activities	Relativities to other process entities

Fig. 4.20 Comparison of tracking object (cf. Fig. 4.16) with event object (cf. Sect. 2.2.2).

4.3.5.1 Composition and Interaction of Control Components

An overview of software components for real-time monitoring and control of manufacturing processes based on tracking of process entities and CEP is depicted in Fig. 4.21. At the bottom, manufacturing resources are employed to execute manufacturing operations, and thereby, manipulate raw materials, products, and the like. Manufacturing data is gathered from these manufacturing resources in (near) real-time (cf. Sect. 4.3.2) using a data collection engine. The data aggregation engine is in charge of integrating this real-time process data with transactional data from enterprise applications (e.g., ERP system). Further, the data aggregation engine manages buffer objects signifying certain manufacturing operations. These buffer objects are streamed to the tracking object manager that creates, updates, and destroys tracking objects. The tracking objects are saved in a relational process database, so that they can be restored from the same. Further, tracking objects are forwarded to the data aggregation engine on creation, update or destruction of a tracking object. The data aggregation engine distributes these tracking objects to registered software components that are interested in tracking objects. There are two types of software components that are fed with tracking objects: (i) visualization clients encompassing dashboards and so forth; and (ii) the online control engine (OCE) based on CEP. The communication among the tracking object manager, data aggregation engine, visualization clients, and the OCE is based on publish-subscribe interfaces. The data aggregation engine is conceptualized as an intermediary or notification service (cf. Sect. 2.2.5). Therefore, the interaction mode of the tracking object manager (i.e., event sender) with the visualization clients and OCE is event based (cf. Sect. 2.2.5).

The tracking objects constitute an event stream that is analyzed by the OCE in (near) real-time. On detection of particular situations, the OCE either creates new events or generates *control objects*, which are forwarded to (i) an OCE administration GUI; and (ii) data aggregation engine for further processing. The composition of a control object is illustrated in Fig. 4.22.

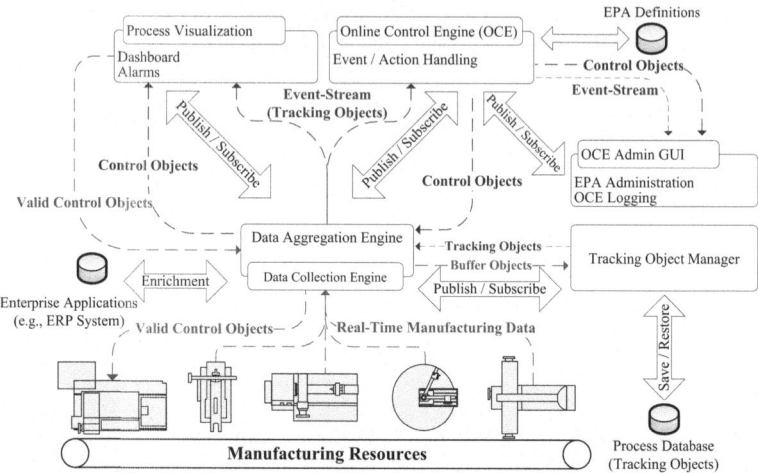

Fig. 4.21 Simplified view of software components for real-time monitoring and control, and event flows among software components.

The property `Type` specifies the further processing of the control object after receipt by the data aggregation engine. The `Priority` is defined as an indicator of the control object's importance or significance. The `EventName` identifies the event pattern that has been detected. This pattern is described with a few words in the property `EventDescription`. The EPL statement that has been formulated to search the event pattern is provided in `EventEPL`. In the case of visualization of an alarm message, the property `UpdateMode` defines how recurrent occurrences of an event pattern have to be visualized (e.g., always add a new alarm message). The instant of time when the control object has been created by the OCE is given in a timestamp called `Created`. The event source can be identified with the properties `ResourceType` and `ResourceName`. In some instances, the receiver of the control object can be confined by setting the property `Role`. The tracking object that has triggered the creation of the control object is specified by `TrackingObjectType` and `TrackingObjectKey`. Finally, a list of messages of type `String` is stored in `MessageList`.

Each message contains placeholders that are marked with leading and trailing tilde symbols (i.e. ~). Each placeholder in the message string is exchanged with the value of a corresponding variable that has been converted into a string. A variable is defined in a control object variable (cf. Fig. 4.22) that is associated with the control object. The property `Name` is an identifier for the corresponding placeholder. The `Type` specifies if the control variable is a measured or planned value, which is stored in `Value`. The data type of the control variable is mentioned in property `DataType`. If the `ControlFlag` is set to the value of 1 and the control variable represents a planned value, then the properties `ControlResourceType`,

`ControlResourceName` and `ControlItemName` are used to manipulate a manufacturing resource. These properties provide sufficient information to write the value of the control object variable into a data item of a data block managed by a PLC. As already mentioned, the control objects are sent to the data aggrega-

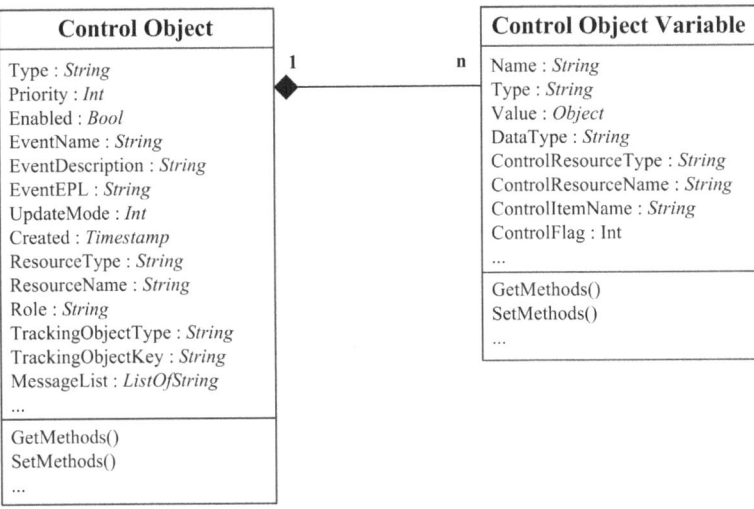

Fig. 4.22 Illustration of control object and associated control object variables, modelled as UML class diagram.

tion engine for further processing. The data aggregation engine analyses the control objects and forwards them to process visualization clients, for instance, to display alarm messages conveyed within the control object. In case of a required control activity (i.e., `ControlFlag` of a control variable has been set to the value of 1), the data aggregation engine invokes a write command, which is based on functionalities of the data collection engine to manipulate data items of a manufacturing resource's PLC. This control activity is illustrated with the path called *valid control object* in the lowermost part of Fig. 4.21. Likewise, a valid control object can also be sent by a process visualization client after confirmation of a control object by an enterprise member.

4.3.5.2 Online Control Engine

An overview on the architecture of the OCE is provided in Fig. 4.23. The main components of the OCE are (i) the CEP engine; (ii) the EPA manager for definition and management of EPAs; and (iii) tracking object pre-processor. The functionality of the OCE is explained according to the illustration in Fig. 4.23.

After start of the OCE, the EPA manager builds EPAs that are primarily composed of (i) a name for identification; (ii) an EPL statement; (iii) a description of the EPL statement; and (iv) a definition of the corresponding control object (cf. step 1). The definitions of these EPAs are stored in an XML configuration file. The EPL statement, which specifies the event processing, is loaded into the CEP engine (cf. step 2). Further, the CEP engine creates a listener for each loaded EPL statement.

A tracking object is received by a tracking object pre-processor (cf. step 3) that converts the tracking object and prepares it for subsequent processing within the CEP engine (cf. step 4). The CEP engine searches the tracking object stream for event patterns employing EPL statements (cf. step 5). An EPL statement expresses temporal, logical and aggregation relationships among various tracking objects. If a certain event pattern has been matched, the registered listener of the corresponding EPL statement is invoked by the CEP engine (cf. step 6). The listener, in turn, checks if the associated EPA contains a control object definition for building a control object (cf. step 7). Finally, a created control object is sent to the data aggregation engine via the interface layer (cf. step 8).

4.3.5.3 Event Processing Agent

The EPAs are central components within CEP (cf. Sect. 2.2.6.2). Luckham has defined an EPA as "an object that monitors an event execution to detect certain patterns of events" [161, 176]. In accordance with the literature (e.g., [67, 161]) the structure of an EPA has been defined as depicted in Fig. 4.24. An EPA has a Name that identifies the EPA. The EPA can be enabled or disabled with the bool property Enabled. Further, an EPA contains event processing statements that define its behavior [67, 161]. There are two methods to define an event processing statement: (i) formulation of an EPL statement as a string; or (ii) instantiation of an EPA object model. The latter can be interactively created by specifying nested objects that represent components of an event processing statement. The EPA object model can be transformed into an equivalent EPL statement by invocation of the method CreateExpression().

The task or meaning of the event processing statement is described in the property Description. The Type specifies if the event processing statement is provided as an EPL statement or an EPA object model. If the EPA is enabled, the resulting EPL statement is loaded by the CEP engine. EPAs can be arranged in EPNs, as elaborated in Sect. 2.2.6.2. Thus, an EPA can be dependent on several other EPAs. Further, the EPLs are loaded sequentially by the CEP engine, thus the property Order is employed to determine the order in which the EPLs have to be loaded.

An EPA is linked with a control object (cf. Fig. 4.22) that is updated and sent to the data aggregation engine after fulfillment of the corresponding EPL statement. Also, an EPA can have a reference to an EPA object model, as depicted in Fig. 4.25 and Fig. 4.26. This model has been adapted from the event processing statement object model developed for the NEsper CEP engine (cf. [63]).

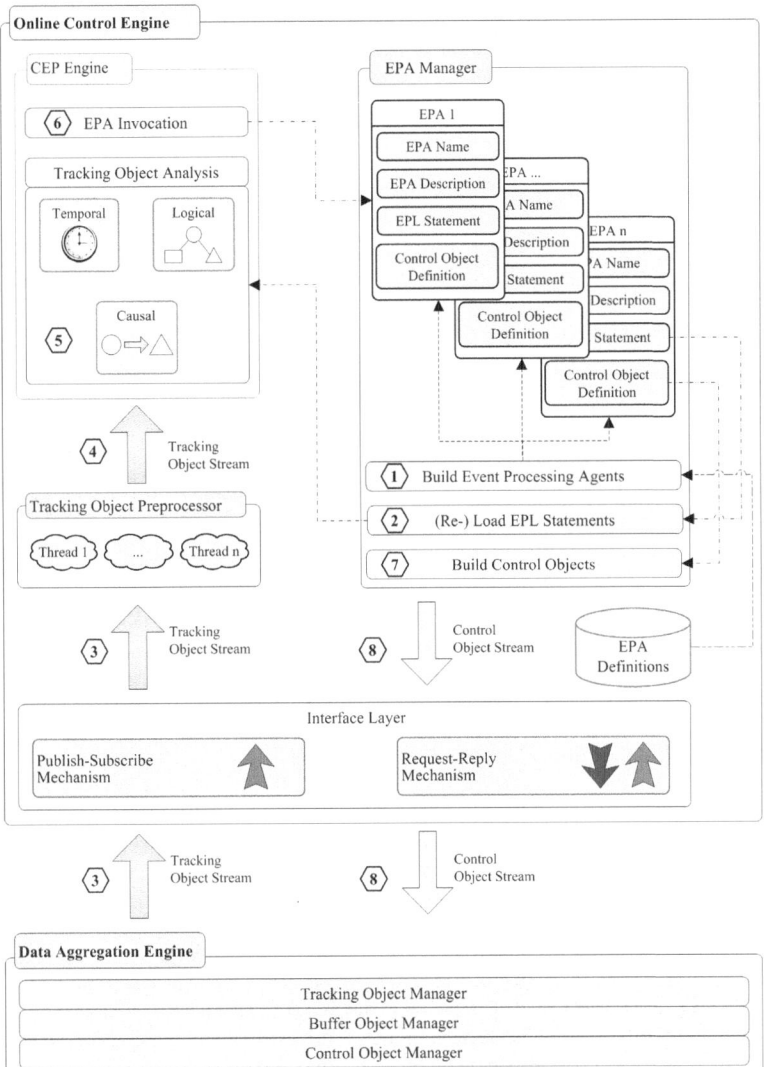

Fig. 4.23 Overview of architecture of online control engine based on CEP.

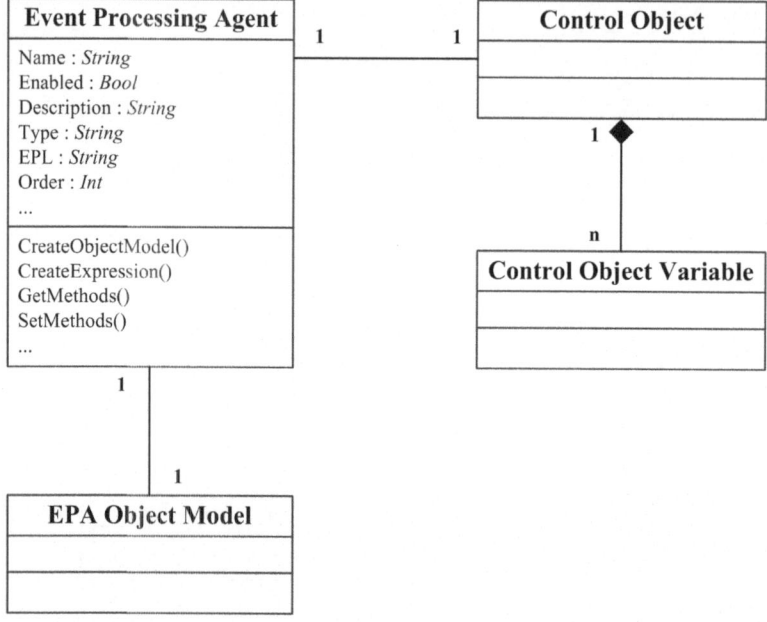

Fig. 4.24 Depiction of an event processing agent (EPA) and relationships to control object and EPA object model, modelled as UML class diagram.

4.3.5.4 Administration of Online Control Engine

The OCE administration GUI is used to provide a means to (i) define and manipulate EPAs; (ii) to visualize the stream of tracking and control objects; and (iii) to monitor the status of the underlying CEP engine. The OCE administration GUI visualizes EPAs in a list and provides functionalities to create, to update and to delete EPAs. The GUI for formulation of an EPL statement capitalizes on the EPA object model.

Further, incoming tracking objects are visualized in a list with their Type and PrimaryKey. On detection of a certain event pattern, which has been formulated as an EPL of an EPA, the EPA's Name is immediately appended and displayed in an EPA history list. This mechanism assists to survey the functioning of EPAs and EPNs by an enterprise member. Further, the status of the CEP engine is shown by mentioning the number of loaded EPL statements, the current frequency of events (e.g., events per hour), the rate of detected event patterns, and the uptime of the CEP engine.

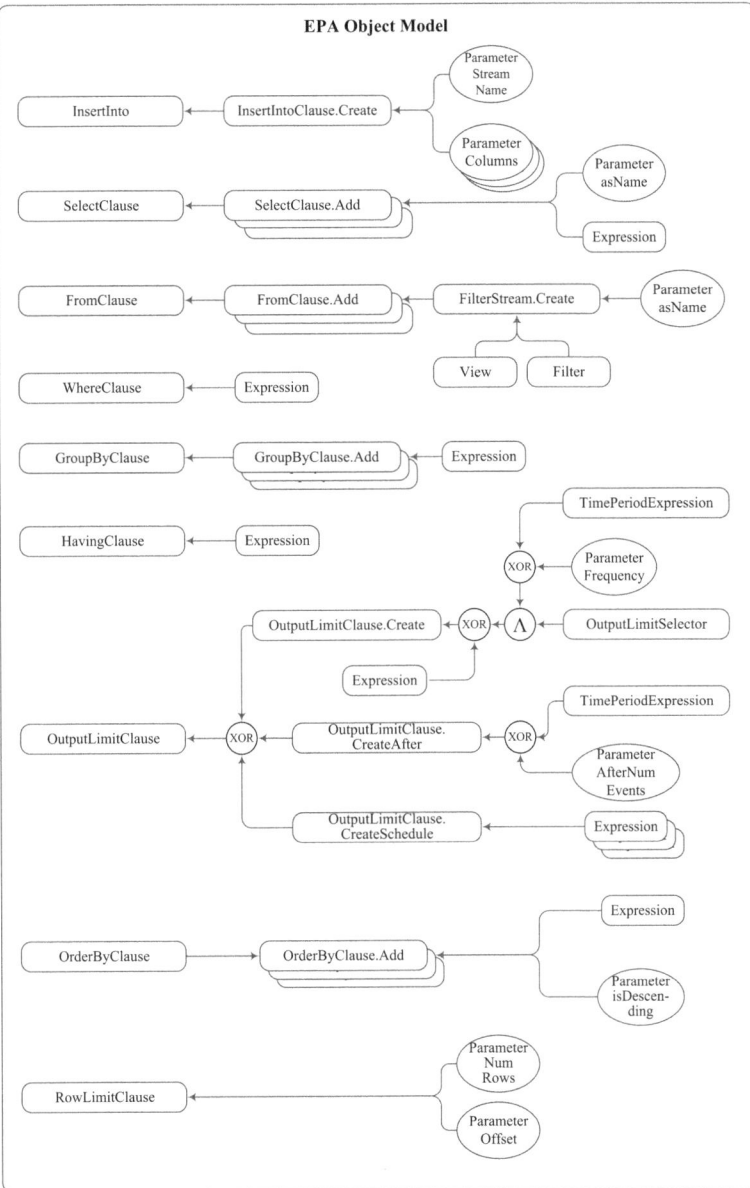

Fig. 4.25 Simplified view of the EPA object model for creation of EPL statements (according to [63]).

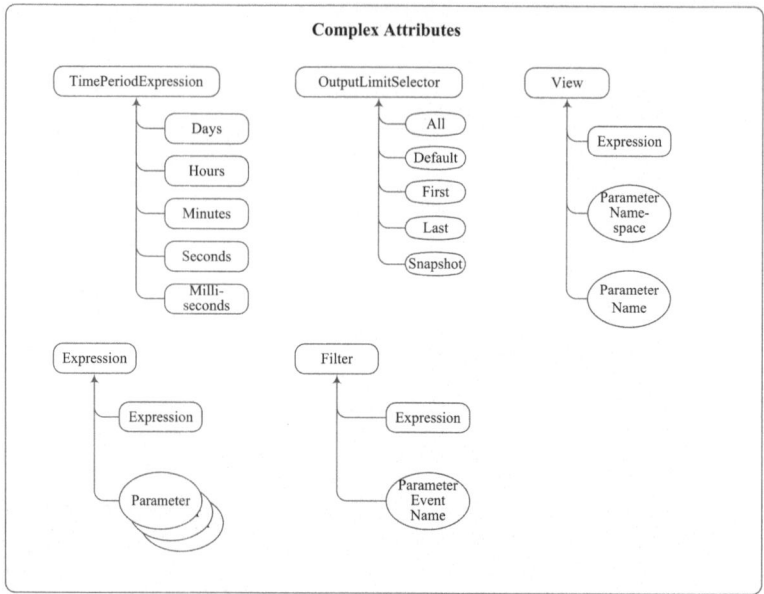

Fig. 4.26 Complex attributes that are used in the EPA object model (according to [63]).

4.3.5.5 Execution of Control Actions

Control objects are transmitted to the data aggregation engine after an event pattern
has been matched. The control objects are analyzed by a control object analyzer
as part of the data aggregation engine. Each control object is processed in a separate thread to achieve high throughput and low latencies. Further, every control
object is stored by the data aggregation engine in the relational process database for
traceability.

There are three possible actions that can be invoked by the control object analyzer: (i) visualization of an alarm message in a visualization client; (ii) dis-
play/sending of advice to an enterprise member that should manipulate a manufacturing resource; and (iii) direct manipulation of a manufacturing resource by writing
process values into a resource's PLC.

The visualization of an alarm message in a visualization client is an elementary
control action. The control objects of Type *AlarmMessage* are just sent to visualization clients where enterprise members that have the same or a more privileged
role than the control object's role are logged in. Decision makers are not interested and are not eligible to receive every control object. For instance, a message
for a manager will not be displayed to an operator on the shop floor. The property
Priority of the control object is employed to display the messages at particular
areas of the visualization client using different colors (e.g., red for high importance).

If a visualization client has been started, the last n relevant messages are requested from the data aggregation engine. Hence, an enterprise member will be informed about alarms that have been raised shortly before the corresponding login process has been initiated.

Control advice can be exposed to an enterprise member who, in turn, accepts or declines the advice. There are two reasons for the application of this approach: (i) the manufacturing resource has no proper communication interface to get manipulated directly; and (ii) the enterprise's policy specifies that only workers/operators shall be responsible to manipulate manufacturing resources. The question of liability is considered and sometimes hinders the realization of an autonomous control approach.

Finally, the control object analyzer can invoke a command, which writes the values of the associated control object variables into data items of resources' PLCs. A control object variable provides sufficient information for the data aggregation engine to parameterize a write command (cf. Fig. 4.22). This control action is performed in soft/firm real-time. Thus, the final decision to manipulate a manufacturing resource's value has to be made by the resource's PLC that is operated in hard real-time. Nevertheless, the presented control approach supplements the resources' control by provision of capabilities to analyze complex event patterns and by the detection of complex process situations that cannot be recognized by a single shop floor resource.

Chapter 5
Implementation of Solution Approach and Evaluation in a Foundry

The aforementioned IT framework has been developed in cooperation with Ohm & Häner Metallwerk GmbH & Co. KG, Olpe, Germany. The enterprise is family-owned, employs more than 400 workers and realizes a turnover of more than 50 million Euros. The working foundry manufactures castings with a total weight starting at 20 grams and goes up to 2000 kilograms (cf. [196]).

State-of-the-art machinery is employed in the new plant[1] in Drolshagen, Germany, to produce components in *small and medium lot sizes*. The company serves more than 400 customers per year, thus has to cope with a *high product variety*. The layout of the new plant primarily follows *product layout*, but partially, also cellular and process layouts are employed. The production encompasses both *continuous* and *discontinuous flow of material*. The outcome of the discrete manufacturing processes are either *bulk goods* (e.g., molten metal) or *piece goods* (e.g., castings, sand cores).

In summary, the manufacturing processes of the considered foundry can be characterized as depicted in Fig. 3.2. Further, according to the characteristics mentioned by Mönch, the established *sand casting processes* in Drolshagen are *complex manufacturing processes* (cf. [189] and Sect. 4.2.2). Following, fundamentals of metal casting in general and sand casting in particular are elaborated.

5.1 Sand Casting

"*Casting* is a process in which molten metal flows by gravity or other force into a mold where it solidifies in the shape of the mold cavity. The term *casting* is also applied to the part that is made by this process" [emphasis in original] [104, 194]. Casting of metals belongs to the group of solidification processes[2] [104].

---

[1] The plant was built in 2008, thus can be considered as one of Europe's most modern foundries for the production of aluminum sand castings (cf. [83]).

[2] Glasswork and processing of polymers are also solidification processes [104].

According to the type of mold, it can be subdivided into permanent-mold casting and expendable-mold casting.

Sand casting is a process in which expendable molds are made out of sand. More than 70 % of all castings are manufactured using sand casting processes [210]. Advantages of sand casting are low cost for equipment, creation of complex component geometries, and its possibility to produce large castings. Sand casting is favored for realization of low-volume production.

5.1.1 Principle of Sand Casting

The principle of sand casting is illustrated in Fig. 5.1. A *mold* is produced out of sand mixtures that are filled into a metal frame called a *flask*. The sand mixtures are compacted applying hydraulic and/or pneumatic pressures. Every mold is composed of a bottom half called a *drag* and a top half called a *cope*. The halves are separated by a *parting line*. *Molten metal* (e.g., aluminum alloy) is poured into the *pouring*

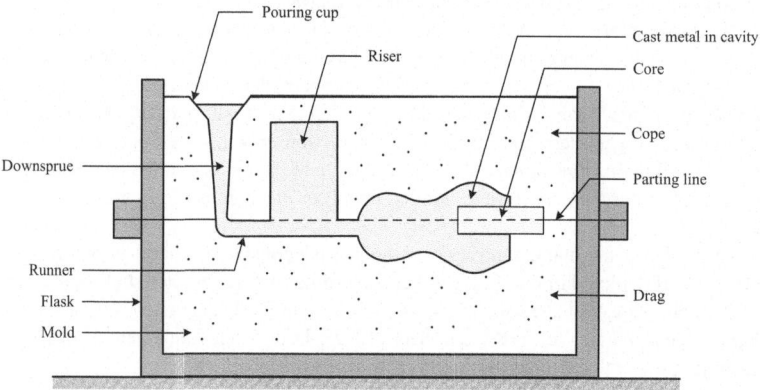

Fig. 5.1 Principle of sand casting process and labelling of important components (adapted from [104]).

cup, flows *down* the *sprue*, passes through the *runner* and finally fills the mold *cavity*. The mold cavity makes up the outer shape of the casting. It is produced by packing sand mixtures around the cope and drag *patterns*, compacting sand mixtures and removing the patterns [104]. The patterns are mounted on plates so that the sand mixtures can be pressed on these patterns during the molding process. The inner shape of the casting can be realized employing *sand cores* that are placed before the drag and cope are assembled.

The molten metal starts to solidify after it has been poured into the mold cavity. Reservoirs of molten metal called risers or feeders are available to compensate for losses because of shrinkage as the molten metal solidifies [49]. The solidified casting is removed by breaking the mold and destroying sand cores.

Numerous types of sand casting defects are known in the practice and described in the literature (cf. [104]). The casting quality depends on sand mixtures, the molding process, melting of metal in a furnace, properties of molten metal (e.g., alloy, temperature), the pouring process, solidification and cooling of the casting, and so forth. The aforementioned factors influence the outcome of the sand casting process. Further, these factors have to be aligned with each other to produce castings with the required quality.

5.1.2 Sand Casting Process

A simplified view of the highly automated sand casting processes of the Ohm & Häner Metallwerk GmbH & Co. KG, Germany, is depicted in Fig. 5.2. Sand core shooting employing cold box processes is performed to produce cores. Cores are necessary if the castings are of a complex shape, thus cannot be shaped with a (simple) drag or cope pattern. Drags and copes are produced on separate molding

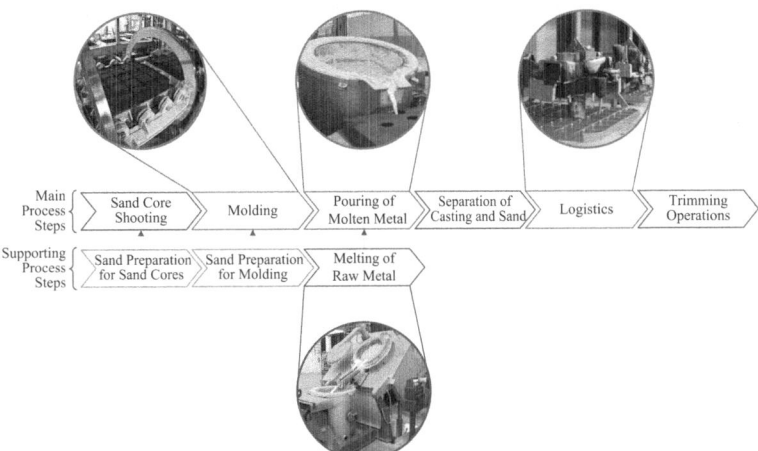

Fig. 5.2 Sand casting process realized in a new plant of the Ohm & Häner Metallwerk GmbH & Co. KG, Drolshagen, Germany (adapted from [101]).

machines, and are assembled shortly before entering one of two available pouring lines. The molding machines have a capacity to produce up to 240 molds per hour. Cores are set into drags to define the inner shape of the casting. Vents are pierced into copes so that gases from molten metal can escape after a mold has been poured.

Aluminum pigs (i.e., small blocks of a certain aluminum alloy that are remelted) and foundry returns (i.e., cut off runners, downsprues, and so forth) are set into a furnace according to an alloy's recipe. The set aluminum is melted until a certain tapping temperature has been reached. Then, the molten metal is poured into a ladle for further processing. A metal treatment process is applied to improve the separation of dross from aluminum and to degass the molten metal. The quality of the molten metal is inspected by thermo analysis (e.g., measurement of density index) and spectral analysis (i.e., chemical analysis of molten metal). If the molten metal is approved for casting it is called a metal lot [49].

The ladle with the metal lot is mounted on a pouring machine, which pours the molten metal into the pouring cup of a mold. Usually, a metal lot is used for many molds. At the end of a pouring line the castings are separated from the mold sand employing a vibrating gutter. This process is called shakeout [49]. The sand is automatically cleaned free of contaminations and recycled, and hence, can be used to prepare further sand mixtures.

Each casting is loaded onto the pendant of a power and free system. The castings are automatically transported to subsequent workstations. First, the castings are parked in cooling lines for a predefined time interval to complete the solidification process. Second, workers use pneumatic hammers to remove sand cores from castings. Third, shot blasting is employed to remove residual sand from the castings' surfaces. Finally, runners, risers and surplus metal are trimmed, and the casting is unloaded from the pendant. A quality inspection is performed for every casting. Further, castings are marked with a (optional) durable data matrix code for product tracking.

Afterward, the castings are transported to the plant in Olpe, where post-processing, like milling and grinding operations, is performed to finalize the castings according to product specifications. In the context of this research work, the manufacturing processes realized in the Drolshagen plant are considered. However, the implemented IT framework can be extended so that the entire value creation chain, encompassing the manufacturing processes in Olpe, can be supported.

5.2 Analysis of Business and Manufacturing Processes

The comprehension of business and manufacturing processes is a prerequisite for the realization of a concrete monitoring and control approach. Thus, a BPR or BPM project precedes the implementation of the envisaged IT framework [95].

The considered manufacturing processes as well as corresponding upstream and downstream business processes have been analyzed. Process owners, like controllers, IT experts and plant managers, have been interviewed. Further, the business and manufacturing processes have been modeled following the ARIS framework. The processes have been modeled in the *control view* of ARIS as EPCs. The focus of the process modeling has been set on the manufacturing processes as these processes were intended for monitoring and control. However, supporting processes, like cost calculation, industrial engineering, and production planning, have also been considered, which are indispensable in providing set points (e.g., product specifications) to control the manufacturing processes.

The considered manufacturing processes are highly automated, thus process data is generated by automation systems and machines in (milli-) seconds. This process data has to be monitored, analyzed for critical process situations, and used to deduce control actions. Therefore, communication interfaces of automation systems and machines have been investigated. Interface descriptions encompassing process data, which can be or has to be delivered by a certain automation system, have been written in tabular form.

Further, the interdependencies between different automation systems and machines have been modeled employing DFDs. The DFDs are organized in a hierarchical manner, starting from an overview DFD for the complete plant/a production line down to detailed DFDs for sub-processes (e.g., melting). A snapshot of the overview DFD of the considered sand casting process is depicted in Fig. 5.3. The interdependency between the molding machine and the six-axis robot for piercing of vent holes is illustrated. Further, the communication interfaces are revealed, which can be employed to acquire process data. Special attention is paid to identify and mention primary/foreign keys, which are exchanged among functions.

5.3 Design of an Enterprise Data Model

The aforementioned DFDs reveal the dynamics of process data creation and interaction of automation systems and machines. The static structure and relations among process data are modeled in an enterprise data model. An extraction of the enterprise data model for the considered manufacturing processes is depicted in Fig. 5.4. The enterprise data model encompasses relevant process entities and describes their relations following the crow's foot notation.

The enterprise data model links process data generated in separated sub-processes (e.g., melting). Further, it relates process data from the shop floor with transactional data from enterprise applications (e.g., ERP system). The implemented ERP system at Ohm & Häner Metallwerk GmbH & Co. KG, Olpe, Germany, is OPTI.V7 that is been developed by the RGU GmbH, Dortmund, Germany. This ERP system addresses the specific needs of foundries. The interface between the MES level and the ERP level is defined in IEC 62264 and encircled in Fig. 5.4. The enterprise data model is a basis to establish forward and backward traceability.

5.4 Implementation of the Framework

The implementation of the framework is based on the aforementioned analysis of the business and manufacturing processes as well as the design of the enterprise

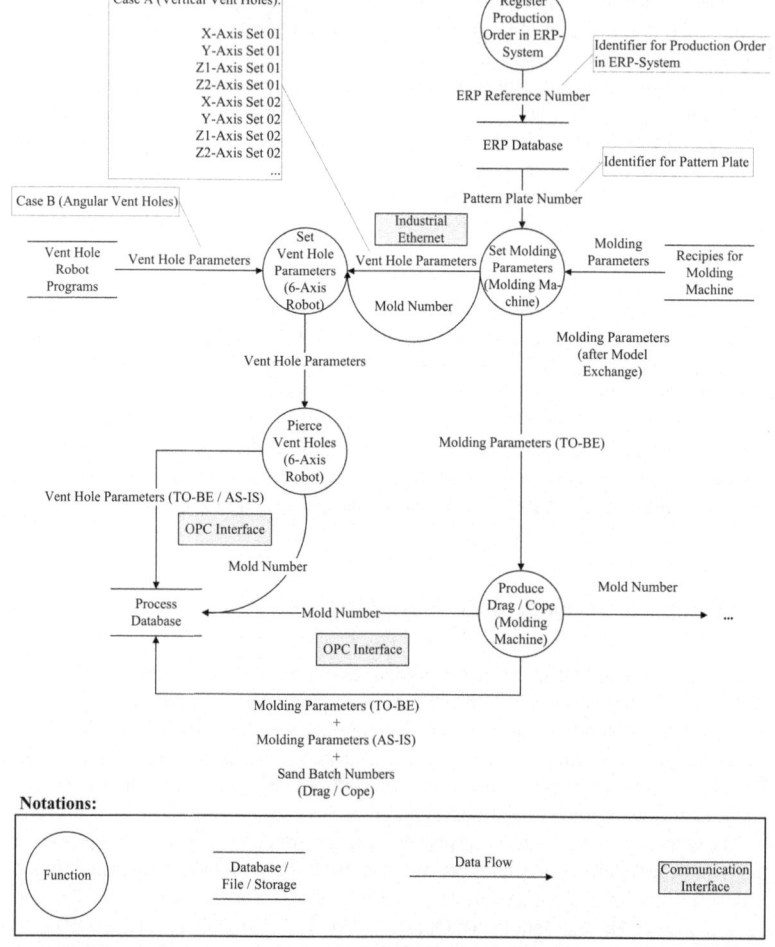

Fig. 5.3 Snapshot of the overview DFD showing the interaction between the molding machine and the robot for piercing vent holes.

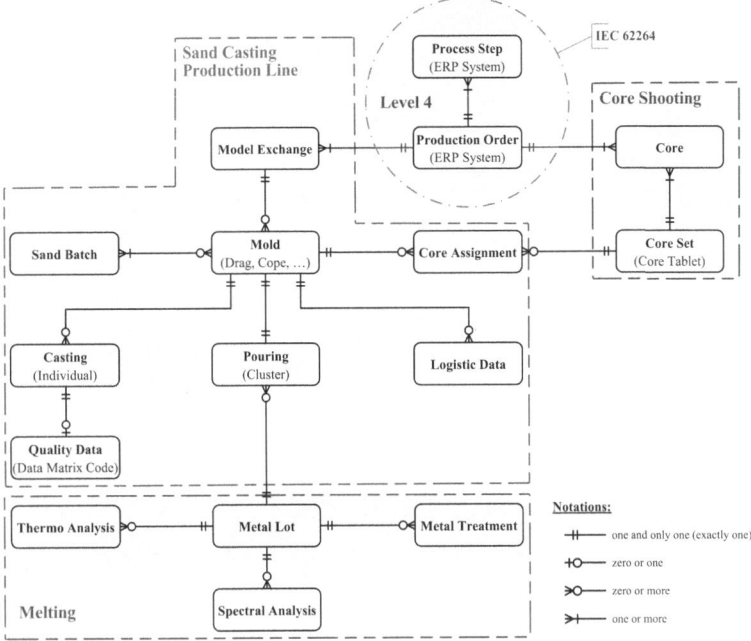

Fig. 5.4 Parts of the enterprise data model covering manufacturing (sub-) processes in the Drol-shagen plant.

data model. The framework has been implemented employing the integrated development environment (IDE) from Microsoft™ called Visual Studio 2010®. The framework capitalizes on the .NET framework 4.0. Most software components are implemented as Windows® services. Such a service can be maintained independently, and further, doesn't require a logged-in user on the server. The software components of the framework are elaborated in subsequent paragraphs.

5.4.1 Process Coverage

The acquisition of process data from automation systems and enterprise applications is a prerequisite for the realization of the envisaged monitoring and control approach (cf. Sect. 4.3.2). The data is stored and managed in a process database, namely Oracle® 10g R2. The database schema is based on the enterprise data model, which has been described in Sect. 5.3. An illustration of the covered manufacturing processes is provided in Fig. 5.5. An estimation of the number of gathered process data items is indicated for each process step. The following manufacturing

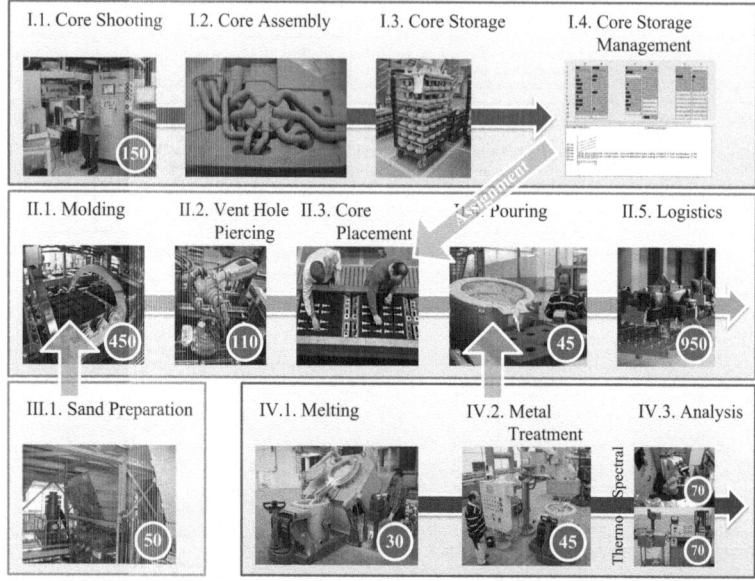

Fig. 5.5 Manufacturing processes covered and supported by the implemented framework.

processes are covered by the developed framework: (i) sand core shooting process inclusive sand core storage management; (ii) sand casting production line encompassing molding machine, pouring machines, and power & free logistics system; (iii) sand preparation process; and (iv) metallurgical process composed of melting, metal treatment, thermo analysis, and spectral analysis. Each of the aforementioned manufacturing processes is integrated in a horizontal direction. Further, assignments among various (sub-) processes are established, as illustrated with arrows in Fig. 5.5. These inter-process relations are described in subsequent paragraphs.

Sand cores are managed by a sand core storage management system, which has been developed at the Department of Business & Information Systems Engineering, University of Siegen. Each sand core is described by a set of process data. Core sets are created during execution of core shooting processes. A core set is a *virtual* collection of sand cores (i.e., process data of the sand cores) and is identifiable by a unique core set number. Each core set is assigned to a *physical* core tablet that, in turn, is part of a core tablet wagon. The physical core tablet is uniquely identified by a barcode, and is registered at a core shooting machine while producing sand cores. Overall, the traceability resolution for sand cores is at a batch level.

The core tablet wagons are parked in a marked storage area composed of numerous spaces on the ground. Later, sand cores are inserted in drags by workers during core placement. Thereby, the core sets are identified by scanning barcodes of the corresponding core tablets. The unique core set numbers of the registered core sets are assigned to the drags, which are identifiable by a unique mold number, wherein the sand cores have been placed.

Sand mixtures are provided by a sand preparation process as sand batches. Each sand batch is identifiable by a unique sand batch number. If a prepared sand mixture is used in a drag or cope, the corresponding sand batch number is assigned to the respective mold. Similarly, a ladle with a metal lot is mounted onto a pouring machine. After pouring of a certain mold, a link is established between the mold (i.e., mold number) and the metal lot (i.e., metal lot number).

5.4.2 Tracking of Process Entities

Process entities are monitored in (near) real-time by a tracking object manager (cf. Sect. 4.3.4). The tracking object manager has been implemented as a separate Windows® service. The tracking objects are defined in an XML file. Which buffer items of a buffer object (type) have to be mapped onto tracking object items of a tracking object (type) are specified. Further, the relativities among various tracking objects are defined. Also, restoration information is provided to restore a tracking object from the process database.

The tracking objects are managed in nested hash tables, thus allow a fast lookup process. The management of the tracking objects is performed in multiple threads for achievement of low latencies. Therefore, the concept of monitors provided by the .NET framework is used to synchronize the access of the aforementioned hast tables and tracking objects. A monitor "controls access to objects by granting a lock for an object to a single thread" [186].

Strict requirements with regard to the robustness of the control components (e.g., tracking object manager) have been formulated. Thus, means have been implemented to ensure that a granted lock for an object will definitely be released by each thread. For instance, constrained execution regions (CER) have been defined that are areas in which "the common language runtime (CLR) is constrained from throwing out-of-band exceptions" (i.e., asynchronous exceptions) [185].

The following tracking object types are managed for monitoring and control of manufacturing processes: (i) *production orders* with transactional data taken from ERP system (i.e., RGU OPTI.V7); (ii) *molds* sub-divided into process data of twelve process steps/manufacturing operations; (iii) *sand batches*; (iv) *metal lots*; (v) *metal treatments*; (vi) *thermo analysis*; (vii) *spectral analysis*; (viii) *molding machine events* denoting disturbances, downtimes, and the like; and (ix) *environmental conditions* incorporating ambient air temperature and humidity.

The tracking objects are forwarded to visualization clients as well as the OCE. A screenshot showing the visualization of process entities is depicted in Fig. 5.6. A *tree of tracking object types* is drawn on the left side. The number of *active* tracking objects is mentioned in square brackets behind each tracking object type name. The *tracking object items* of the selected tracking object are displayed right beside the tracking object tree. The tracking items are grouped according to their *process activities*. Further, a tracking object has *relativities* to other tracking objects. A status of the server components and an overview on the sand casting production line is displayed in the rightmost part of the screenshot.

5.4.3 *Process Analysis using Historical Process Data*

Analysis of the historical process data can be performed to achieve insights with regard to the manufacturing processes' efficiency. The implemented framework offers different types of manufacturing process analysis, especially, performance, molten metal, product, and quality analysis. Although, in most instances, the aforementioned analyses are not employed for real-time monitoring and control of manufacturing processes, the retrospective analysis can reveal process disturbances and insufficiencies. Further, these analyses facilitate the identification of control-related knowledge (cf. [98]).

5.4.3.1 Performance Analysis

The performance analysis is performed for the sand casting production line. The performance analysis is achieved both in (near) real-time and as an aggregation of performance values over a period of production dates. An online performance analysis for of the sand casting production line is depicted in Fig. 5.7. The molding machine is the main production resource of the sand casting production line. The actual *takt time*[3] of the sand casting production line can be determined by measuring the takt time of the molding machine. The acquired process data of each molding operation contains a creation timestamp that indicates when the molding operation has been performed. The elapsed time between two produced molds can be measured by calculating the difference between consecutive creation timestamps. The quotient of the aggregated elapsed times and the number of produced molds represents the measured takt time. The actual achieved takt time can be contrasted with the average planned takt time (cf. Fig. 5.7). The quotient derived based on these takt times is an indicator of the *performance* of the sand casting production line. The performance is limited to 100 %, even though the actual achieved takt time is much lower than the planned takt time.

The elapsed times are classified according to their duration. A time classification is depicted in Fig. 5.7. A threshold has been defined in order to determine at which elapsed time value a process inefficiency can be assumed. Thus, the elapsed times can be either assigned to a class "Production Time OK" or "Production Time Not OK" These time classes are used to derive the *availability* of the sand casting production line. The availability is the "Production Time OK" divided by the com-

---

[3] Takt time is understood as the time that is required to manufacture a product. Often, the term cycle time is used interchangeably.

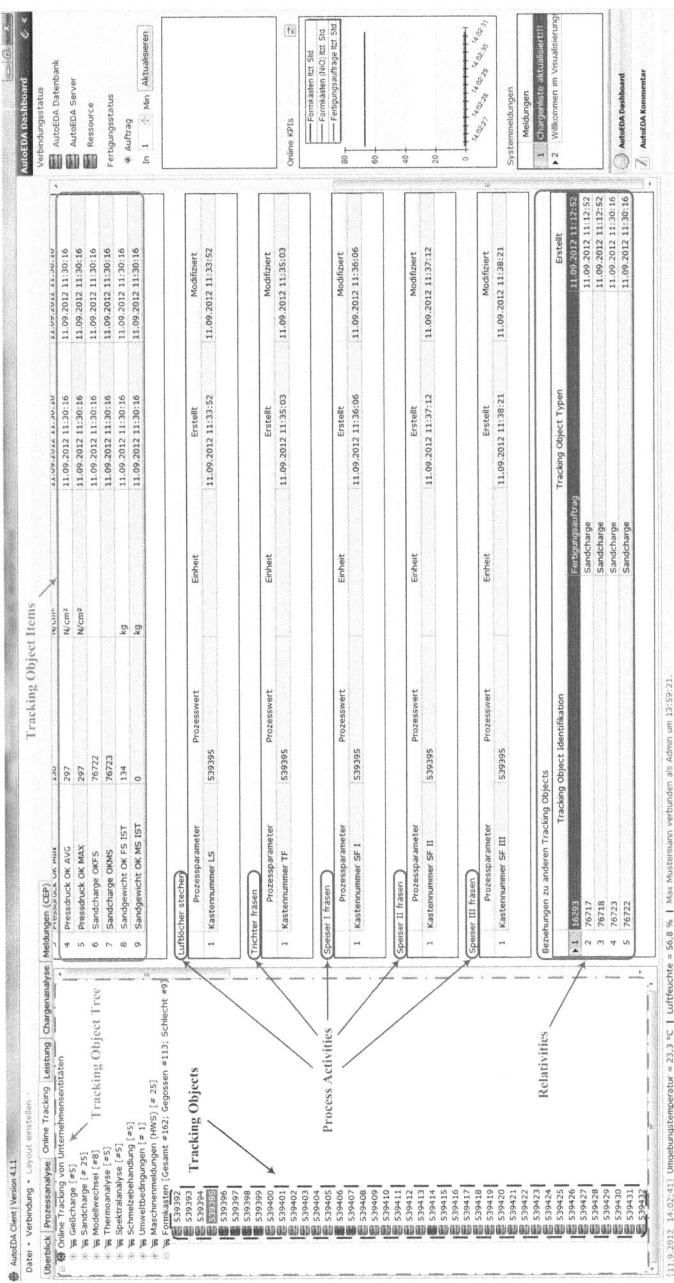

Fig. 5.6 Screenshot of visualization client showing online tracking of various process entities (in German).

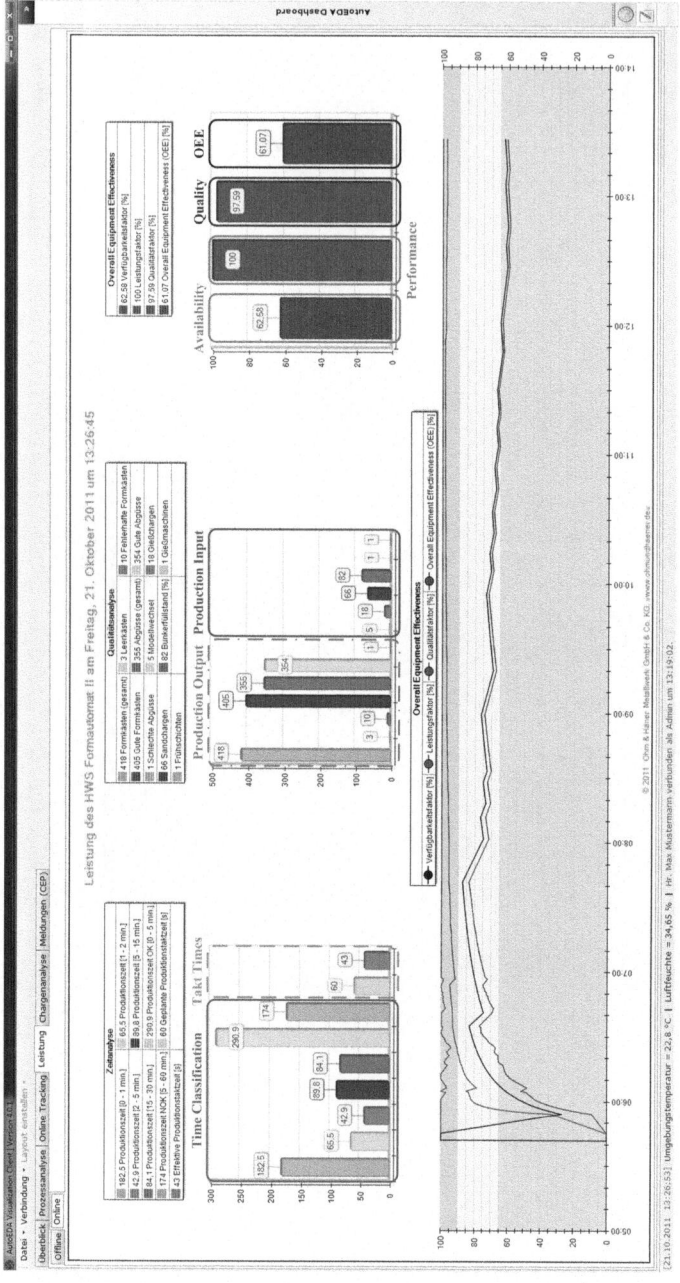

Fig. 5.7 Performance analysis of the considered sand casting production line (in German).

pletely available production time[4] (i.e., "Production Time OK" plus "Production Time Not OK").

The production output is measured by counting the number of produced molds and castings. Also, the quality information of the molds and the castings (cluster) are captured. This quality information is employed to derive a *quality* indicator for the sand casting production line.

In summary, the overall equipment effectiveness (OEE) is calculated, which is the product of the aforementioned OEE components, i.e., availability, quality, and performance, of the sand casting production line. The OEE is a compact KPI for evaluation of the effectiveness of the established transformation processes. The following rating for OEE and its components has been defined for the considered sand casting production line: (i) low OEE/OEE component ranges from 0 % to 65 %; (ii) a medium OEE/OEE component ranges from 65 % to 90 %; and (iii) an high OEE/OEE component starts from 90 % and higher. For each of these ratings, a certain color has been defined to visualize the OEE components and the OEE (cf. Fig. 5.7).

5.4.3.2 Molten Metal Analysis

A molten metal analysis has been developed that covers the metallurgical manufacturing processes (cf. Sects. 5.1.2 and 5.4.1). The visualization of the molten metal analysis is illustrated in Fig. 5.8. Molten metal batches can be searched by providing parameters, like time range, customer name, molten lot number, alloy number, and so forth. After performing a process step for a molten metal batch (e.g., spectral analysis), corresponding process parameters are automatically analyzed and a status for the molten metal batch is determined. The status of a molten metal batch is derived from the status of both thermo and spectral analyses. If a molten metal batch adheres to the aluminum alloy's intended elemental composition and sticks to the required physical properties, like modification, grain refinement, and density index, then its status is set to OK (i.e., approved for pouring). If either the spectral or thermo analysis reveals limit violations, then the status of the molten metal batch is set to NOK (i.e., pouring of the molten metal batch is prohibited). Every status change is documented in an event protocol.

The searched molten metal batches are visualized in a list, and different colors are used to signal the status of the molten metal batches (e.g., green for OK and red for NOK). The process values of the selected molten metal batch are displayed, and limit violations are highlighted in red. An editing and commentary process can be employed to approve a molten metal batch, even though it contains limit violations. Therefore, an authorized worker has to select a comment to justify the approval of the violated molten metal batch. The name of the authorized worker is recorded for traceability reasons. An approved molten metal batch is called a metal lot (cf. [49]), and can be used for the pouring of molds.

---

[4] Times for planned maintenance and planned breaks are excluded from the available production time. A discussion of time classes for calculation of MES-related KPIs is presented in [89].

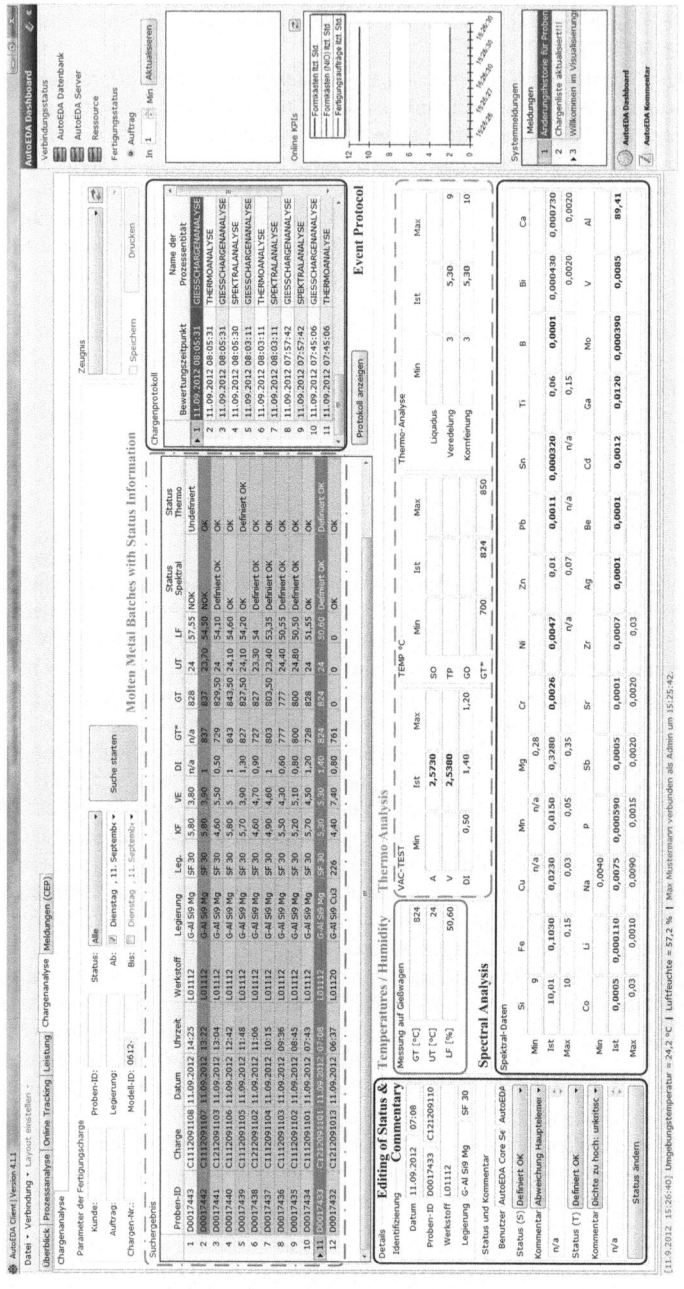

Fig. 5.8 Analysis of molten metal batches encompassing thermo and spectral analyses, and displaying history of molten metal batch status with an event protocol (in German).

5.4.3.3 Product Analysis

Product analysis is employed to investigate the production efficiency with regard to a certain product. The considered foundry has a high product mix, and often, products have to be produced with low volumes. Thus, an analysis of actual takt times of a certain product for several production dates can deliver insights, e.g., about (i) the quality of production schedules; and (ii) effectiveness of CIP.

An example of a takt time analysis as part of product analysis is shown in Fig. 5.9 and Fig. 5.10. The measured takt time is contrasted with the planned takt time. The example illustrates the improvement of the measured takt time. At the first production date a takt time of 143 s (i.e., bad run) was achieved. After taking corrective measures, the measured takt time could be reduced to 62 s (i.e., best run). The average takt time is calculated incorporating the corresponding relative lot sizes. In the example, an average takt time of 75 s has been calculated. Further, each production date can be analyzed in more depth by clicking on one of the visualized production dates[5]. A Gantt chart summarizes the requested production date by displaying the (historic) production schedule. Compact information about a product is depicted on the axis of the ordinates. Horizontal bars are shown for each product. Each horizontal bar represents a production slot. An annotation contains details with regard to a production slot (e.g., measured takt time versus planned takt time). The horizontal bars can be commented by authorized plant managers, workers, and the like. The number of assigned comments is indicated for each product with a bracketed, red number. Further, information is provided showing in which pouring line the produced molds have been poured.

The retrospective analysis of measured takt times and historic production schedules assists a plant manager in identifying favorable/unfavorable sequences of production slots. This knowledge can be used to improve the production scheduling and initiate an appropriate CIP.

5.4.3.4 Quality Analysis

The implemented framework supports the survey of (intermediate) product quality. Usually, in an ERP system, quality feedback is provided as aggregated values for a finished production order. As such, it is not possible for a quality assurance representative to trace out-of-spec products. This shortcoming is mitigated by the developed framework by capturing quality feedback for individual input factors and (intermediate) products.

A quality certificate can be exported based on the collected quality feedback, as depicted in Fig. 5.11. The quality certificate summarizes the feedback concerning required reworking activities and casting rejections. This feedback has been provided by workers after inspection of individual castings. The aggregated values are shown by bar diagrams and pie charts. Further, an authorized user (e.g., plant

---

[5] This functionality is sometimes also referred to as *drill down functionality*.

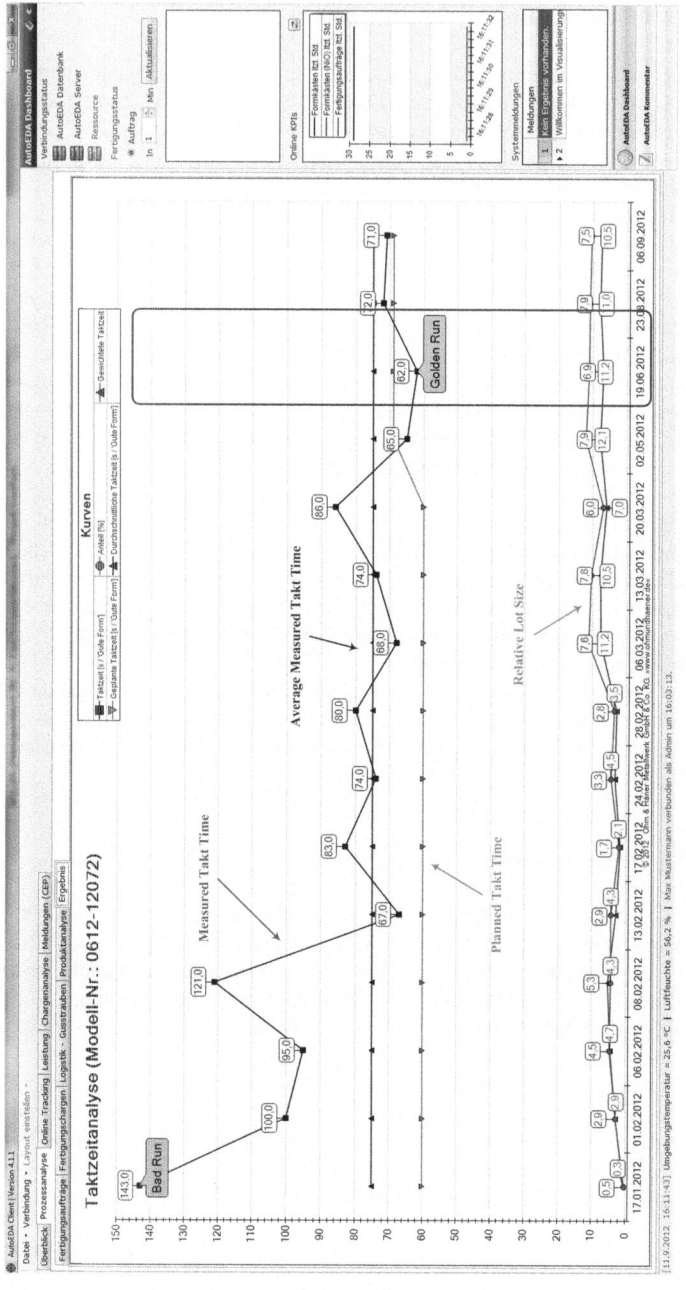

Fig. 5.9 Visualization of takt times for different production dates as part of product analysis (in German).

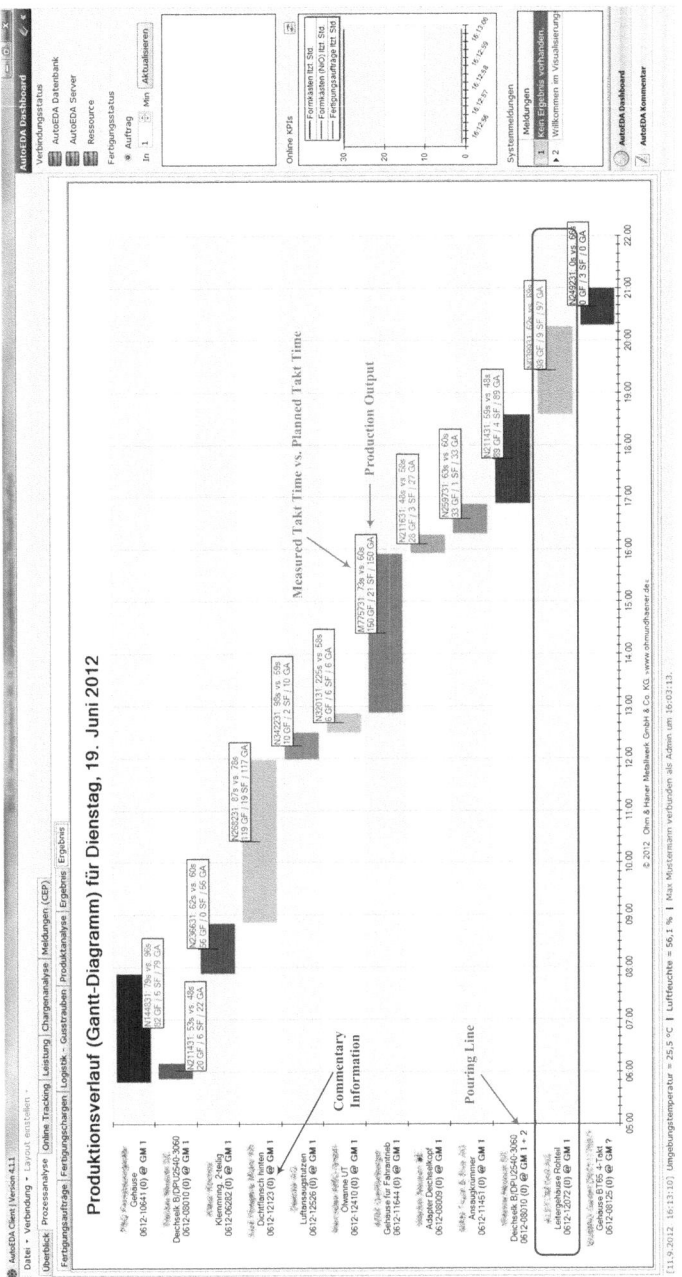

Fig. 5.10 Gantt chart showing a historic schedule of production orders and their condensed performance information (in German).

manager, quality assurance representative) is able to dig down to obtain quality information of each individual casting. The availability of quality information for

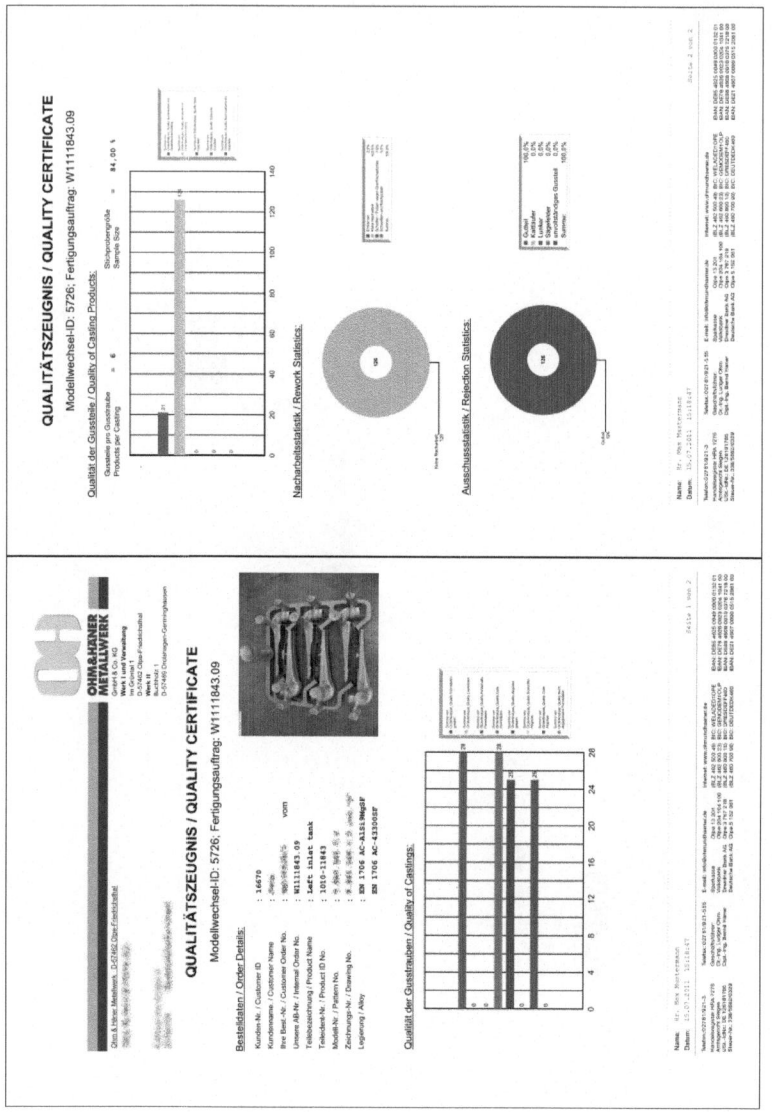

Fig. 5.11 Export of quality certificate for an individual production order, and visualization of rework and rejection statistics.

each individual casting supports the analysis of causes of defects. This root cause analysis can be accompanied by a KDD process, as described in Sect. 4.2.3.

5.4.4 Real-Time Control of Sand Casting Processes

The elaborated process analysis is performed retrospectively. It provides helpful insights with regard to process disturbances, production scheduling, quality failures, and so forth. This knowledge can be employed for real-time monitoring and control of manufacturing processes. Therefore, an authorized user, most likely a plant manager, can formulate EPAs and link them to EPNs.

A screenshot of the OCE administration GUI is depicted in Fig. 5.12. EPAs can be defined using a graphical editor. An EPL statement can be represented as a tree, which is built in accordance with the EPA object model (cf. Sect. 4.3.5.3). This tree can be compiled into an EPL statement. The EPL statement is loaded into the CEP engine and is employed for real-time monitoring. In addition, a control object definition can be provided by the user. For instance, an alarm message can be specified that is raised if the corresponding EPL statement conditions are fulfilled. The user can monitor the event stream, which is composed of tracking objects, and can observe the invoked EPAs, as illustrated in Fig. 5.13. If an EPA with an alarm message definition has been invoked, the alarm message is highlighted in an EPA history list. Further, the status of the OCE can be investigated by the user. For instance, the event and EPA throughputs are visualized in contrast with the overall memory consumption and the central processing unit (CPU) usage. In the following subsections, a selection of real-time monitoring and control examples is presented. The elaborated event patterns have been implemented for the Drolshagen plant. Further, the event patterns have been presented and discussed with domain experts (cf. [181]). The events and event patterns are modeled in accordance with an adapted version of the business event modeling notation (cf. [51] and Sect. 2.2.6.4). The graphical elements for event modeling are depicted in the lower part of Fig. 5.14.

5.4.4.1 Static Threshold Violation

The detection of a static threshold violation is a common monitoring task. An example is provided in the upper part of Fig. 5.14. The tracking object of a sand batch encompasses information about the filling of four sand bunkers. These sand bunkers must be sufficiently filled to provide sand to the molding machine.

After receipt of a new sand batch, the event properties that describe the bunker fillings are analyzed. If a bunker filling value is lower than 30, a control object is initiated. This control object can be interpreted as an output event, which contains an alarm message. The alarm message is forwarded to the data aggregation engine that, in turn, forwards the alarm message to registered visualization clients.

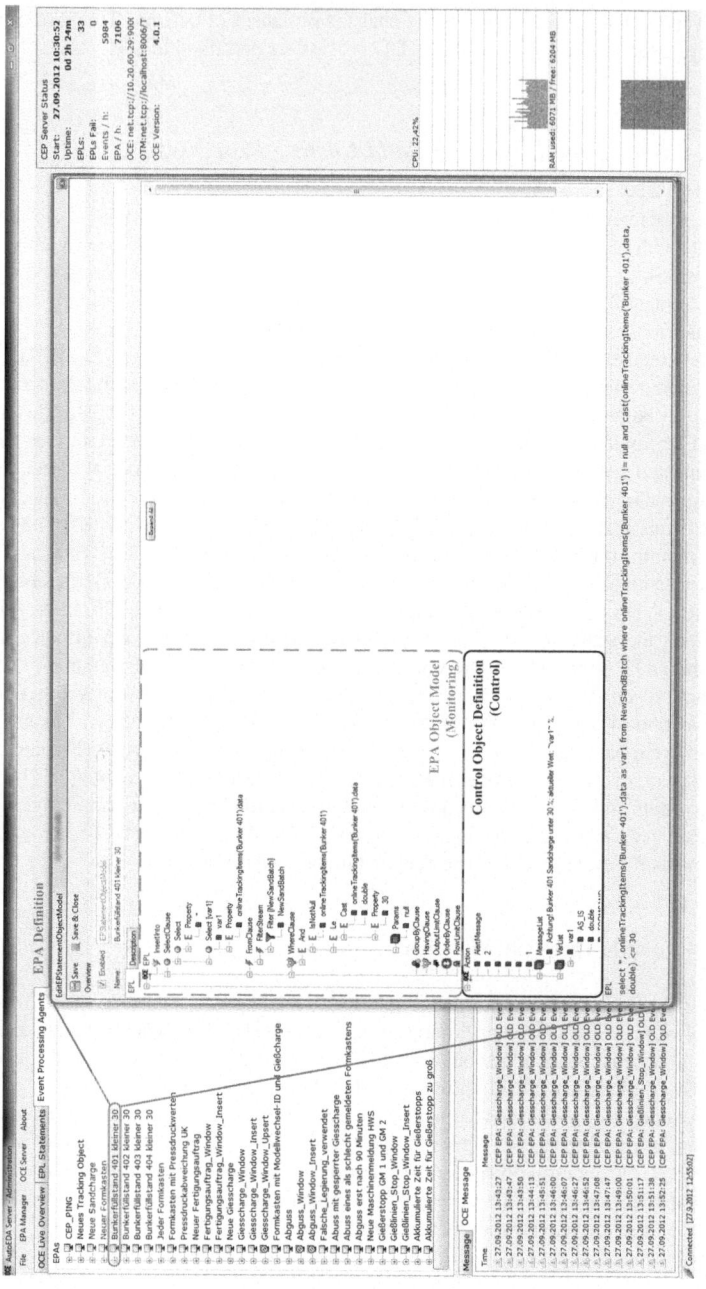

Fig. 5.12 OCE administration GUI displaying a list of EPAs, and showing the definition of an EPA for detection of an empty sand bunker.

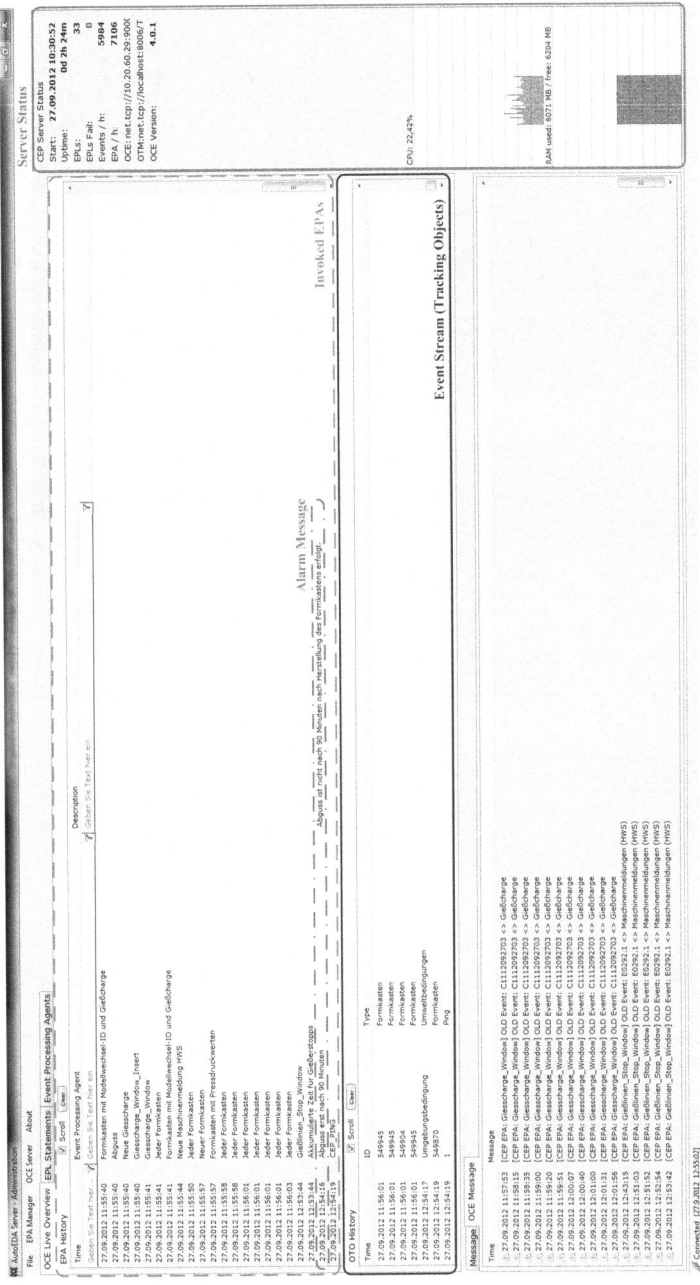

Fig. 5.13 OCE administration GUI showing the incoming event stream composed of tracking objects and the invoked EPAs.

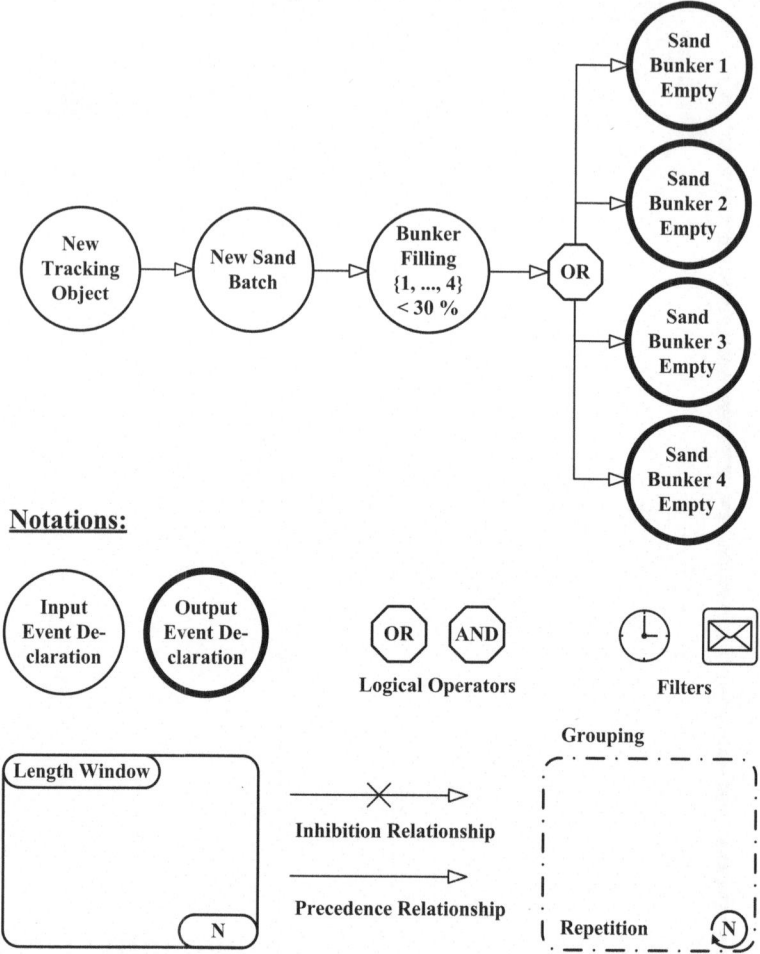

Fig. 5.14 Detection of a violation of a static threshold, and explanation of elements of the BEMN.

5.4.4.2 Product-Based Control

In most instances, it is required to manufacture products in adherence to product specifications. The product specifications have been defined in upstream processes. An example for a product-based control is illustrated in Fig. 5.15. A new order tracking object is created after mounting of pattern plates and a subsequent exchange of those pattern plates at the molding machine. The order tracking object entails properties of the castings, like alloy, pouring temperature, density index, and so

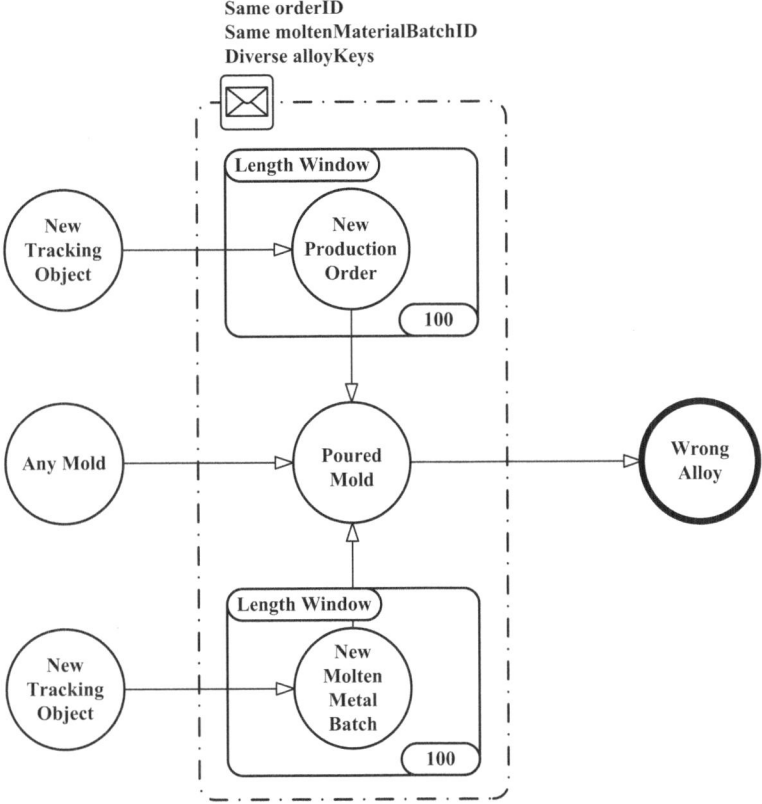

Fig. 5.15 Detection of non-adherence to product specifications as an example of a product-based control.

forth. These properties have been defined in a production planning process, thus are loaded from the ERP system. The last 100 order tracking objects are managed in a length window (cf. Sect. 2.2.6.3).

Similarly, a new molten metal batch tracking object is created after a furnace is tapped. A molten metal batch tracking object contains information about the alloy, tapping temperature, and the like. The last 100 molten metal batch tracking objects are managed in a length window.

A mold tracking object has relativities to a production order tracking object as well as to a molten metal batch tracking object after it has been poured. If the poured mold has been poured with the wrong alloy, the alloy information in the production order tracking object differs from the alloy information mentioned in the molten metal batch tracking object. In that case, a control object is created that conveys a corresponding alarm message.

Noteworthy, the illustrated example reveals the capability of the control approach to link transactional data from the enterprise control level with process data from the shop floor. Thus, the control approach is based on the vertical (information) integration of an enterprise.

5.4.4.3 Control with Temporal Conditions

In some instances, it is indispensable to incorporate temporal conditions into monitoring and control of manufacturing processes. An illustration of such an event pattern is depicted in Fig. 5.16. A mold has to be poured no later than 90 minutes after its production. A mold that is older than 90 minutes is likely to be dry and condensation water might have been built on top of inserted cooling irons. Both effects probably cause defects on a casting's surface. If a pouring event does not

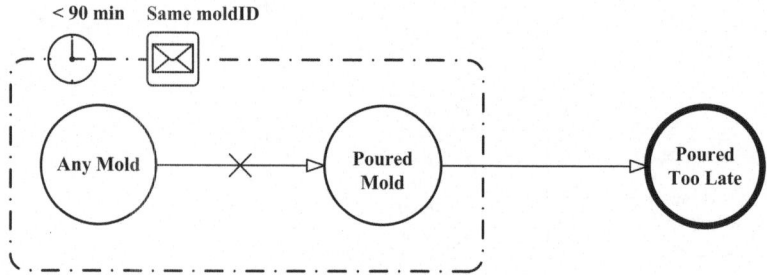

Fig. 5.16 Control incorporating temporal conditions for detection of a non-event.

occur within 90 minutes after the corresponding mold has been produced, a control object is which contains an alarm message is created. Subsequently, a plant manager, a master or a responsible worker can proactively react (i.e., prohibition of the pouring action). The capability of the control approach to investigate temporal conditions, and even, to react on non-events differentiates it from most approaches that are based on RBS.

5.4.4.4 Environmental Conditions

Aluminum sand casting uses batches (e.g., sand, molten metal) as input materials. The sand casting processes are influenced by changing environmental conditions. Therefore, detection of threshold violations is mandatory. The humidity has a negative impact on the molten metal batches. If the humidity is too high, the metal treatment procedure has to be adapted to mitigate this negative impact. However, an alarm message that indicates a humidity threshold violation should only be raised after 10 consecutive occurrences of the threshold violation. The event pattern that realizes this control behavior is depicted in Fig. 5.17.

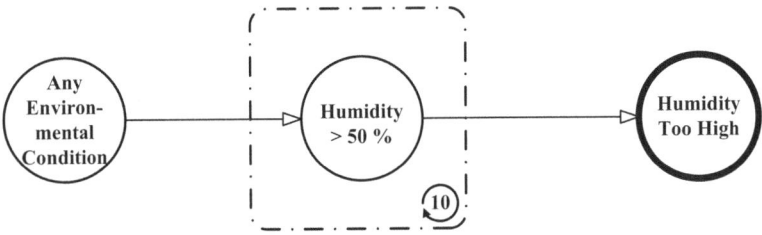

Fig. 5.17 Control of environmental conditions, illustrated with a repeated violation of a humidity threshold.

5.5 Evaluation of Manufacturing Process Improvements

The presented research work follows the methodological approach of the German design-oriented ISR (cf. Sect. 1.3). The insufficient realization of the RTE vision in manufacturing, i.e., especially, the insufficient vertical integration of a manufacturing enterprise and the inadequate real-time monitoring and control of manufacturing processes, has been identified as a relevant research gap/challenge (cf. Sect. 2.1). An attempt, i.e., a framework based on EDA and CEP, has been made to solve the above problems.

The framework has been developed, implemented and tested in a foundry. This research approach is called *prototyping* in the German design-oriented ISR [19], which is also considered as an *evaluation approach* in the design-oriented ISR [214]. Although the evaluation against reality is of high significance, it is rarely applied in the design-oriented ISR [214]. Riege et al. suppose three reasons for this situation: (i) evaluation is a relatively new discipline in the design-oriented ISR (e.g., best practices are missing); (ii) evaluation against reality is time consuming; and (iii) there is no access to/availability of an enterprise that serves as an evaluation partner [214]. As the IT artifact (i.e., implemented framework) is used in a foundry, it is *evaluated against reality*. In that sense, the presented results are significant for both researchers and practitioners that are coping with the RTE in manufacturing and its related challenges.

The implemented IT framework can be seen as a significant contribution to the realization of an RTE in manufacturing. The elaborated methodology provides guidelines and means for the implementation of an RTE. The developed IT framework can be employed to implement an MES. In line with Thiel (cf. [248]), the benefits of an MES, and thus, also of the presented IT framework, are summarized in Fig. 5.18.

A benefit can be classified according to its degree of fulfillment. Further, in some instances, the improvement can be measured. However, most improvements cannot be measured exactly. The following paragraphs discuss the non-measurable and measurable improvements.

Benefit	**Classification**	**Measurability**
Process Transparency	Very High	Non-Measurable
Time Saving / Enhanced Takt Time	Very High	Measurable
Reduction of Indirect Value Creation	High	Measurable
Improved Customer Care Service	High	Non-Measurable
Improved Quality	Very High	Non-Measurable
Early Warning System, Real-Time Cost Control	High	Non-Measurable
Increase of Efficiency of Labor	High	Non-Measurable
Adherence to Guidelines (e.g., VDI 5600)	Very High	Non-Measurable

Fig. 5.18 Benefits of the IT framework for establishment of an RTE in manufacturing (adapted from [248]).

5.5.1 Non-Measurable Improvements

First, the developed framework enhances the transparency of the manufacturing processes. This process transparency is provided retrospectively as well as in (near) real-time. At any time, a plant manager has precise information about the status of the manufacturing processes, manufacturing resources, production orders, products, and so forth.

The *process transparency* can be seen as a means to investigate process disturbances, analyze quality failures, evaluate production schedules, validate measures taken as part of the overall CIP, and the like. In addition, these effects influence support processes, like the *customer care service*. A customer can be informed about the progress of a certain production order. For instance, a necessary bottleneck in production can be detected at the earliest, and thus, a postponement of the delivery date can be communicated on time.

Further, the *quality* of the manufacturing processes and the manufactured products can be improved. The actual process data is constantly compared with planned data, like product specifications. Deviations are detected in (near) real-time and production parameters can be adapted to achieve higher qualities. Root causes for quality failures can be determined, and consequently, appropriate actions to mitigate these quality failures can be taken.

The CEP technology is also used to establish an *escalation management*. Manufacturing process disturbances and deviations are recognized in (near) real-time. Therefore, required (re-) actions can be taken immediately after occurrence of a disturbance/deviation. Thus, the response times can be reduced. Thereby, CEP facilitates the processing of *complex* event patterns, thus enables the detection of *complex* process situations. In addition, the framework is the basis for the realization of a *real-time cost control*[6].

The *efficiency of labor* is increased as workers are confronted with the actual efficiency of the performed manufacturing processes. As a consequence, workers are motivated to improve the manufacturing processes' efficiency and actively take part in CIP. Finally, the framework enhances the *adherence to guidelines*, like VDI 5600 and VDA 6.1.

Tasks five to eight of VDI 5600 (cf. Sect. 3.3.2.1) are realized with the implemented framework. The implemented tasks encompass the most often requested features of an MES, i.e., data acquisition and performance analysis (cf. [271]). Similarly, the control functions three to seven and eleven that are mentioned by ISA 95 (cf. Sect. 3.3.2.3) are covered by the implemented framework.

5.5.2 Measurable Improvements

In addition to the non-measurable improvements, a reduction of actual takt times can be measured after introduction of the IT framework. The *average measured takt times* for the implemented sand casting production line in the Drolshagen plant are shown in Fig. 5.19. A significant improvement of the average measured takt time can be identified after the introduction of the product analysis, Gantt charts, and OCE based on CEP. From March 2011 to January 2012, a decrease in the *average measured takt time* of 17.6 % was achieved. At the same time, the *number of castings* (i.e., production output) could be increased by 17.1 % and the *required production time* (i.e., production input) was reduced by 3.3 %.

The aforementioned time savings and the significant increase of the manufacturing processes' efficiency is the result of numerous improvement measures. Thus, the implemented IT framework facilitates the CIPs. The reworking of the manufacturing processes, organizational structures, and so forth, are fostered by the availability of precise information about manufacturing processes in (near) real-time. Hence, the IT framework optimally unfolds its benefits if it is aligned with the overall management of the manufacturing enterprise.

---

[6] The research work of Karadgi is concerned with an approach for real-time cost accounting (cf. [136]).

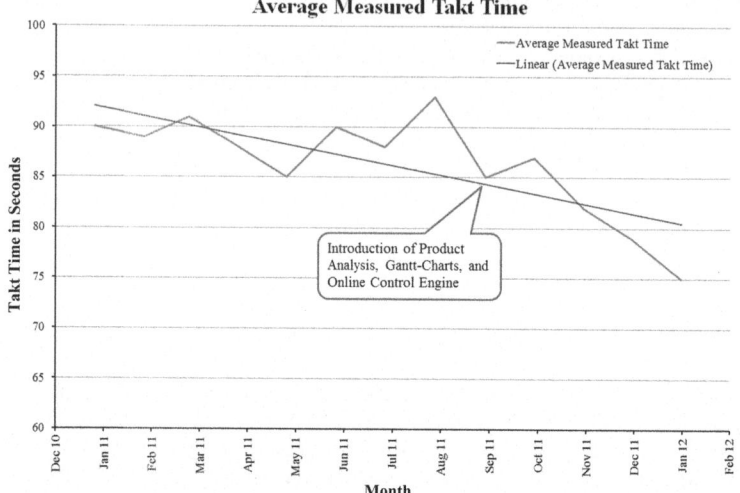

Fig. 5.19 Average measured takt time for the sand casting production line in the Drolshagen plant.

5.5.3 Comparison with Requirements

The above discussion indicates that the implemented framework supports decision making related to production, as discussed for the *management perspective* in Sect. 3.2. The framework has positive effects upon the quality, quantity, time, and cost of production. The return on investment for the implemented framework is less than one year.

Moreover, the framework realizes multiple closed-loop controls that are conceptualized in the *engineering perspective* (cf. Sects. 3.3.1 and 3.3.2). Although not all tasks/functions that are listed in VDI 5600 and ISA 95, have been implemented so far, the framework is a basis for the realization of a complete MES.

The requirements from a *computer science perspective* (cf. Sect. 3.4) are fulfilled by the implemented framework. Firstly, the processing of events is performed in *firm/soft real-time*. Secondly, the real-time analysis of events, which are created during execution of manufacturing processes, incorporates the relevant *business context* (e.g., technical parameters are linked with corresponding order information). Thirdly, the presented real-time monitoring and control approach are capable of *processing complex events*. Finally, flexibility and adaptability of the implemented framework are assured by *separating* the *event processing logic* from the *application logic* (e.g., possibility to add, edit, and delete EPL statements).

Chapter 6
Summary, Conclusions and Future Work

The necessity to (re-) act on internal and external events in (near) real-time has been formulated as part of the RTE vision. This vision requires an enterprise with fully integrated, automated and individualized value creation processes. With respect to manufacturing enterprises, the following problem areas could be identified in this research work: (i) standardization of shop floor interfaces to establish a vertically integrated enterprise; (ii) conceptualization of real-time monitoring and control concepts based on EDA and CEP; and (iii) their implementation in the realm of MES.

The seamless horizontal and vertical integration is a prerequisite for an immediate transfer of information from its POC to an appropriate POA. Further, the RTE requires measures to select and analyze relevant information, thus avoids an indiscriminant flood of data. Finally, a continual alignment of planned and actual process execution is envisioned by an RTE.

Overall, the realization of the aforementioned RTE in manufacturing requires the consideration of management, computer science, and engineering perspectives. The management community's view of a manufacturing enterprise and its production management is focused on topmost enterprise levels (i.e., strategic and tactical enterprise levels). Further, from a management perspective, ERP is considered as a suitable approach for support of value creation processes.

However, a deeper and broader integration of the manufacturing level (i.e., shop floor) has been researched by the engineering community. Their research and development activities have led to various standards and implementations of MES. Recently, MES has been mentioned as a means to establish an RTE in manufacturing. In addition, control engineering has provided principles for the realization of multiple closed-loop controls (i.e., feedback controls). At this point, it is noteworthy that, for instance, cybernetics has been considered as a theoretical foundation and principle of the RTE. Nevertheless, major issues remain open with respect to the interface to shop floor resources.

The RTE was formulated by Gartner as a vision, thus does not explicitly define with which IT it has to be implemented. Nevertheless, the RTE's behavior has been described as being event-driven. Therefore, EDA and CEP have been identified as enablers of an RTE. These paradigms enable IT systems, which adhere to the following system requirements of an RTE to be built: (i) agility; (ii) timeliness; and (iii) availability of information. Unfortunately, so far, CEP has been employed only for financial and administrative processes within and across enterprises. Concepts and approaches for real-time monitoring and control of *manufacturing processes* that capitalize on CEP are rare. Also, the liaison of MES and CEP requires further attention in research.

A framework for the realization of an RTE in manufacturing has been conceptualized, implemented, and evaluated in this research work. This framework has been designed for manufacturing enterprises, which (i) produce tangible products; (ii) are SMEs; and (iii) are part of the German industry. Thus, the manufacturing enterprises and their manufacturing processes, which are supported by the developed framework, have been characterized. Thereby, the relevance of the manufacturing sector for the German economy has been discussed.

Before delving into the details of the developed framework, a related work on intelligent monitoring and control approaches, like MAS, HMS, and so forth, has been provided. Some ideas and concepts of these approaches could be transferred to the presented event-driven framework. The framework consists of two parts: (i) a process model that describes the methodology for the introduction of the IT architecture; and (ii) the design and implementation of the IT architecture for the realization of an RTE in manufacturing.

The process model entails (i) analysis and (re-) design of business and manufacturing processes; (ii) design of an enterprise data model and modeling of data flows between (IT) systems; (iii) identification of control-related knowledge employing, for instance, a KDD process and interviews with domain experts; and (iv) the use of this knowledge for monitoring and control of manufacturing processes.

The last two steps of the process model assume the existence of integrated process data, event-driven monitoring and control mechanisms. Therefore, an IT architecture has been developed that capitalizes on EDA and CEP. Process data is acquired from manufacturing resources and integrated with transactional data from enterprise applications. This functionality can be used to realize backward and forward traceability. Further, a real-time tracking of enterprise entities, like production orders, products, batches, and so forth, has been developed.

The stream of tracking objects can be interpreted as an event stream. This event stream is analyzed by a CEP engine to detect critical process situations and deduce appropriate (re-) actions. Noteworthy, the tracking objects contain actual process data as well as planned data from enterprise systems (e.g., ERP system). Therefore, the established control approach is based on an integrated enterprise, which is demanded by an RTE. Also, the remaining RTE principles, i.e., automation of decision-making processes and the immediate availability of information at the POA, are fulfilled by the developed IT architecture.

The process model and the event-driven framework have been implemented for a foundry. The framework is employed to monitor and control highly automated aluminum sand casting processes of the Ohm & Häner Metallwerk GmbH, Olpe, Germany, in their new plant in Drolshagen, Germany. The current release of the implemented framework has led to a significant reduction of the average measured takt time by around 18 % and a return on investment in less than one year.

So far, the presented framework is limited to manufacturing enterprises and manufacturing processes that have been characterized in Sect. 3.1. Further, the vision of an RTE entails other aspects of an enterprise like personnel, sales, and so forth. Up to now, the framework has not dealt with these aspects. Consequently, the extension of the framework to cover these aspects can be part of future work. In addition, the framework can be extended to implement more tasks/functions of MES.

The extension of the framework to other manufacturing processes of the considered foundry has been envisaged. The mechanical processing (e.g., turning, milling) especially has to be implemented to complete the coverage of the overall value creation process. In addition, control actions have to be implemented that directly influence manufacturing resources. An example is the blocking of a pouring machine, if the wrong alloy is used or the (measured) pouring temperature is expected to not adhere to defined product specifications. Also, the transfer of the control approach to manufacturing resources has been planned. This development fits into the recently promoted research pertaining to Industry 4.0 and CPS.

References

[1] van Aalst, W.d.: Process Mining: Discovery, Conformance and Enhancement of Business Processes. Springer, Heidelberg and New York (2011)

[2] Aalst, W.M.P., Hofstede, A.H.M., Weske, M.: Business process management: A survey. In: W.d. van Aalst, A.t. Hofstede, M. Weske (eds.) Business Process Management: International Conference on Business Process Management (BPM), pp. 1–12. Springer, Berlin (2003)

[3] Abele, E., Reinhart, G.: Zukunft der Produktion: Herausforderungen, Forschungsfelder, Chancen. Carl Hanser, München (2011)

[4] Abolhassan, F.: Das Real-Time Enterprise - Eine Einordnung. In: A.W. Scheer, F. Abolhassan, W. Bosch (eds.) Real-Time Enterprise, pp. 1–7. Springer, Berlin (2003)

[5] Abrahams, P.: Simple infrastructure for EDA processor agent (2005). URL http://www.it-director.com/technology/content.php?cid=7983

[6] Agha, G.A.: Actors: A Model of Concurrent Computation in Distributed Systems, *Technical Report*, vol. 844 (1985)

[7] Aho, A.V., Lam, M.S., Sethi, R., Ullman, J.D.: Compilers: Principles, Techniques, & Tools, 2 edn. Pearson Addison Wesley, Boston (2007)

[8] Aier, S., Schelp, J.: How to preserve agility in service oriented architectures – an explorative analysis. Enterprise Modelling And Information Systems Architectures 5(2), 21–37 (2010)

[9] Alavi, M., Leidner, D.E.: Review: Knowledge management and knowledge management systems: Conceptual foundations and research issues. MIS Quarterly 25(1), 107–136 (2001)

[10] Allen, J.F.: Towards a general theory of action and time. Artificial Intelligence 23(2), 123–154 (1984)

[11] Alt, R., Cäsar, M., Leser, F., Österle, H., Puschmann, T., Reichmayr, C.: Architektur des Echtzeitunternehmens. In: R. Alt, H. Österle (eds.) Real-time Business, pp. 20–52. Springer, Berlin (2004)

[12] Alt, R., Österle, H. (eds.): Real-time Business: Lösungen, Bausteine und Potenziale des Business Networking. Springer, Berlin (2004)

[13] Alves, A., Arkin, A., Askary, S., Barreto, C.e.a.: Web Services Business Process Execution Language Version 2.0: OASIS Standard (April 2007). URL http://docs.oasis-open.org/wsbpel/2.0/OS/wsbpel-v2.0-OS.html

[14] Ammon, R.v., Emmersberger, C., Greiner, T., Springer, F., Wolff, C.: Event-driven business process management. In: Fast Abstract, Second International Conference on Distributed Event-Based Systems (DEBS 2008). Rome and Italy (2008)

[15] Anil Kumar, S., Suresh, N.: Production and Operations Management: (With Skill Development, Caselets and Cases), 2 edn. New Age International (P) Ltd. Publishers, New Delhi (2008). URL http://site.ebrary.com/lib/academiccompletetitles/home.action

[16] Babulak, E., Wang, M.: Discrete event simulation. In: A. Goti (ed.) Discrete Event Simulation: State of the Art, pp. 1–10. Sciyo (2010)

[17] Bartlett, D.: CORBA notification services (2001). URL http://www.ibm.com/developerworks/library/co-cjct7/

[18] Bayer, M.: Alles Realtime oder was? Computerwoche (11-12), 14–17 (2012)

[19] Becker, J., Holten, R., Knackstedt, R., Niehaves, B.: Forschungsmethodische Positionierung in der Wirtschaftsinformatik – epistemologische, ontologische und linguistische Leitfragen – (2003)

[20] Bernus, P., Nemes, L.: The contribution of the generalised enterprise reference architecture and methodology to consensus in enterprise integration. In: K. Kosanke, J. Nell (eds.) Proceedings of International Conference on Enterprise Integration Modeling Technology (ICEIMT), pp. 175–189. Springer, Heidelberg (1997)

[21] Bertalanffy, L.v.: General System Theory: Foundations, Development, Applications. Braziller, New York (1968)

[22] Bierer, W., Liappas, I.: Strategieentwicklung im Real-Time Enterprise (RTE). In: A.W. Scheer, F. Abolhassan, W. Bosch (eds.) Real-Time Enterprise, pp. 11–23. Springer, Berlin (2003)

[23] van Biljon, S.S.: Role of access to 'real-time' information in the survival of enterprises. Ph.D. thesis, University of Stellenbosch, Stellenbosch and South Africa (2004)

[24] Birrell, A.D., Nelson, B.J.: Implementing remote procedure calls. ACM Transactions on Computer Systems 2(1), 39–59 (1984)

[25] Blair, E.: latency (2006). URL http://searchcio-midmarket.techtarget.com/definition/latency

[26] Blood, S.: The RTE: It starts with early warnings. In: B. Kuhlin, H. Thielmann (eds.) The Practical Real Time Enterprise, pp. 351–357. Springer, Berlin and London (2005)

[27] Bloom, N., Genakos, C., Sadun, R., van Reenen, J.: Management practices across firms and countries (2011). URL http://www.hbs.edu/research/pdf/12-052.pdf

[28] Böhmann, T., Leimeister, J.M.: Integration von Produkt und Dienstleistung - Hybride Wertschöpfung

(2010). URL http://www.uni-goettingen.de/de/
integration-von-produkt-und-dienstleistung---hy\
bride-wertschoepfung/107847.html

[29] Bohn, H., Bobek, A., Golatowski, F.: SIRENA - service infrastructure for
real-time embedded networked devices: A service oriented framework for
different domains. In: P. Lorenz, P. Dini, D. Magoni, A. Mellouk (eds.)
Proceedings of International Conference on Networking, International Conference on Systems, International Conference on Mobile Communications
and Learning Technologies (ICONS 2006), pp. 43–48. IEEE, Piscataway and
NJ and USA (2006)

[30] Bosch, W.: Auf dem Weg zum Real-Time Enterprise. In: A.W. Scheer,
F. Abolhassan, W. Bosch (eds.) Real-Time Enterprise, pp. 173–180. Springer,
Berlin (2003)

[31] Brandl, D., Cleal, G., Gifford, C.e.a.: Data architecture for MOM: The manufacturing master data approach: White paper #37 (2010)

[32] Bruns, R., Dunkel, J.: Event-Driven Architecture: Softwarearchitektur für
ereignisgesteuerte Geschäftsprozesse. Springer, Berlin (2010)

[33] van Brussel, H., Wyns, J., Valckenaers, P., Bongaerts, L., Peeters, P.: Reference architecture for holonic manufacturing systems: PROSA. Computers in
Industry 37(3), 255–276 (1998)

[34] Bry, F., Eckert, M., Etzion, O., Paschke, A., Riecke, J.: Event processing
language tutorial: Presentation, 3rd ACM international conference on distributed event-based systems (DEBS 2009), Nashville, TN, USA, July 6-9,
209 (2009). URL http://www.slideshare.net/opher.etzion/
debs2009-event-processing-languages-tutorial

[35] Buffa, E.S.: Modern Production Management: Managing the Operations
Function, 5 edn. Wiley/Hamilton, Santa Barbara and Calif (1977)

[36] Buffa, E.S.: Modern Production/Operations Management, 6 edn. Wiley, New
York (1980)

[37] Bullard, V., Vambenepe, W.: Web services distributed management: Management using web services (MUWS 1.1) part 1 (2006). URL http://docs.
oasis-open.org/wsdm/wsdm-muws1-1.1-spec-os-01.htm

[38] Cammert, M., Heinz, C., Krämer, J., Riemenschneider, T., Schwarzkopf, M.,
Seeger, B., Zeiss, A.: Stream processing in production-to-business software.
In: L. Liu, A. Reuter, K.Y. Whang, J. Zhang (eds.) Proceedings of 22nd
International Conference on Data Engineering (ICDE 2006), p. 168. IEEE
(2006)

[39] Chakravarthy, S., Jiang, Q.: Stream Data Processing: A Quality Of Service
Perspective : Modeling, Scheduling, Load Shedding, and Complex Event
Processing. Springer, New York (2009)

[40] Chandramouli, B., Goldstein, J., Barga, R., Riedewald, M., Santos, I.: Accurate latency estimation in a distributed event processing system. In: Proceedings of 27th IEEE International Conference on Data Engineering (ICDE), pp.
255–266 (2011)

[41] Chandy, K.M., Schulte, W.R.: Event Processing: Designing IT Systems for Agile Companies. McGraw Hill, New York (2010)

[42] Chen, D., Doumeingts, G.: The GRAI-CIM Reference Model, Architecture and Methodology. In: P. Bernus, L. Nemes, T.J. Williams (eds.) Architectures for Enterprise Integration, pp. 102–126. Chapman & Hall, London (1996)

[43] Chinnici, R., Moreau, J.J., Ryman, A., Weerawarana, S.: Web Services Description Language (WSDL) Version 2.0 Part 1: Core Language (June 2007). URL http://www.w3.org/TR/wsdl20/

[44] Choudhary, A.K., Harding, J.A., Tiwari, M.K.: Data mining in manufacturing: A review based on the kind of knowledge. Journal of Intelligent Manufacturing 20(5), 501–521 (2009)

[45] Clark, C.: The Conditions of Economic Progress. Macmillan, London (1940)

[46] Clark, T., Barn, B.S.: A common basis for modelling service-oriented and event-driven architecture. In: Proceedings of the 5th India Software Engineering Conference (ISEC 2012), pp. 23–32 (2012)

[47] Colombo, A., Schoop, R., Neubert, R.: An agent-based intelligent control platform for industrial holonic manufacturing systems. IEEE Transactions on Industrial Electronics 53(1), 322–337 (2006)

[48] Corsten, H., Gössinger, R.: Produktionswirtschaft: Einführung in das industrielle Produktionsmanagement, 12 edn. Oldenbourg, München (2009)

[49] Courtesy American Foundry Society: Glossary of metalcasting terms (2012). URL http://www.xeniafoundry.com/images/FoundryTerms.pdf

[50] Dash, M., Liu, H., Yao, J.: Dimensionality reduction of unsupervised data. In: Proceedings of 9th International Conference on Tools with Artificial Intelligence (ICTAI '97), pp. 532–539. IEEE (1997)

[51] Decker, G., Großkopf, A., Barros, A.P.: A graphical notation for modeling complex events in business processes. In: Proceedings of 11th IEEE International Enterprise Distributed Object Computing Conference (EDOC), pp. 27–36 (2007)

[52] DeRemer, F., Kron, H.: Programming-in-the large versus programming-in-the-small. ACM SIGPLAN Notices 10(6), 114–121 (1975)

[53] Di Wang, Rundensteiner, E.A., Ellison III, R.T.: Active complex event processing over event streams. Proceedings of the VLDB Endowment (PVLDB) 4(10), 634–645 (2011)

[54] Die Bundesregierung: Industrie 4.0 (2012). URL http://www.hightech-strategie.de/de/2676.php

[55] Diltis, D., Boyd, N., Whorms, H.: The evolution of control architectures for automated manufacturing systems. Journal of Manufacturing Systems 10(1), 79–93 (1991)

[56] Dorp, C.A.v.: Reference-Data Modelling for Tracking and Tracing. Wageningen (2004)

[57] Dove, R.: Knowledge management, response ability, and the agile enterprise. Journal of Knowledge Management 3(1), 18–35 (1999)

[58] Driscoll, D., Mensch, A.: Devices profile for web services version 1.1: OA-SIS standard (2009). URL http://docs.oasis-open.org/ws-dd/dpws/wsdd-dpws-1.1-spec.html

[59] Driver, M.: Java and .NET: You Can't Pick a Favorite Child: Keynote Presentation, ESRI Developer Summit, Palm Springs, California (2007). URL http://proceedings.esri.com/library/userconf/devsummit07/papers/keynote_presentation-mark_driver_gartner.pdf

[60] Drobik, A., Raskino, M., Flint, D., Austin, T., MacDonald, N., McGee, K.: The Gartner definition of real-time enterprise (October 2002)

[61] Eckert, M., Bry, F., Brodt, S., Poppe, O., Hausmann, S.: A CEP babelfish: Languages for complex event processing and querying surveyed. In: J. Kacprzyk, S. Helmer, A. Poulovassilis, F. Xhafa (eds.) Studies in Computational Intelligence, vol. 347, pp. 47–70. Springer, Berlin and Heidelberg (2011)

[62] Enste, U., Mahnke, W.: OPC Unified Architecture. at - Automatisierungstechnik 59(7), 397–404 (2011)

[63] EsperTech: Reference documentation. version: 4.4.0 (2011)

[64] EsperTech: FAQ (2012). URL http://esper.codehaus.org/tutorials/faq_esper/faq.html

[65] ESPRIT Consortium AMICE: CIMOSA: open system architecture for CIM, 2 edn. Springer, Berlin (1993)

[66] etalis: etalis - event-driven transaction logic inference system (s.a.). URL http://code.google.com/p/etalis/

[67] Etzion, O., Niblett, P.: Event Processing in Action. Manning, Stamford and CT and USA (2011)

[68] European Commission: COMMISSION RECOMMENDATION of 6 may 2003 concerning the definition of micro, small and medium-sized enterprises: EU recommendation 2003/361 (2003). URL http://eur-lex.europa.eu/LexUriServ/LexUriServ.do?uri=OJ:L:2003:124:0036:0041:EN:PDF

[69] Faison, T.: Event-Based Programming: Taking Events to the Limit. Springer, New York (2006)

[70] Fandel, G., Fistek, A., Stütz, S.: Produktionsmanagement, 2 edn. Springer, Berlin (2011)

[71] Fayyad, U., Piatetsky-Shapiro, G., Smyth, P.: From data mining to knowledge discovery in databases. AI Magazine 17(3), 37–54 (1996)

[72] Federal Ministry of Economics and Technology (BMWi): Renaissance der Industrie und die Rolle der Industriepolitik: Bedeutung des Verarbeitenden Gewerbes und Standortpolitik der Bundesregierung (2008). URL http://www.bmwi.de/BMWi/Redaktion/PDF/Publikationen/renaissance-der-industrie,property=pdf,bereich=bmwi,sprache=de,rwb=true.pdf

[73] Federal Statistical Office: Export, Import, Globalisierung: Deutscher Außenhandel und Welthandel, 1990 bis 2008 (2010).

URL   https://www.destatis.de/DE/Publikationen/
Thematisch/Aussenhandel/Gesamtentwicklung/
AussenhandelWelthandel5510006099004.pdf?__blob=
publicationFile

[74] Federal Statistical Office: Deutsche Wirtschaft: 4. Quartal 2011 (2012).
URL   https://www.destatis.de/DE/Publikationen/
Thematisch/VolkswirtschaftlicheGesamtrechnungen/
DeutscheWirtschaftQuartal.pdf?__blob=
publicationFile

[75] Ferstl, O.K., Sinz, E.J.: Grundlagen der Wirtschaftsinformatik, 6 edn. Oldenbourg, München (2008)

[76] Finkenzeller, K.: RFID Handbook: Fundamentals and Applications in Con-
tactless Smart Cards and Identification, 2 edn. Wiley, Chichester and England
and Hoboken and NJ (2003)

[77] Fleisch, B.D.: Distributed shared memory in a loosely coupled distributed
system. In: J.J. Garcia-Luna-Aceves (ed.) Frontiers in Computer Communications Technology, pp. 317–327. Association for Computing Machinery,
New York and NY and USA (1987)

[78] Fleisch, E., Österle, H.: Auf dem Weg zum Echtzeit-Unternehmen. In:
R. Alt, H. Österle (eds.) Real-time Business, pp. 3–17. Springer, Berlin
(2004)

[79] Forgy, C.: Rete: A fast algorithm for the many pattern/many object pattern
match problem. Artificial Intelligence **19**(1), 17–37 (1982)

[80] Foundation, O.: OPC Unified Architecture – advantages and possibilities of
use for manufacturers and users of automation, IT or MES products (2012).
URL http://www.opcfoundation.org/DownloadFile.aspx/
Brochures/OPC-UA-Overview.pdf?RI=801

[81] Fourastié, J.: Die große Hoffnung des zwanzigsten Jahrhunderts, 3 edn.
Bund-Verlag, Köln-Deutz (1954)

[82] Frank, U.: Towards a pluralistic conception of research methods in information systems research (2006). URL http://hdl.handle.net/
10419/58156

[83] Franken, M.: Welcome to 21st century. Giesserei **97**, 90–99 (2010)

[84] Freitas, A.A.: A survey of evolutionary algorithms for data mining and
knowledge discovery. In: A. Ghosh (ed.) Advances in Evolutionary Computing, pp. 819–845. Springer, Berlin (2003)

[85] Gamma, E., Helm, R., Johnson, R., Vlissides, J.: Design Patterns CD: Elements of Reusable Object-Oriented Software. Addison Wesley (1998)

[86] Garcia-Molina, H., Salem, K.: Main memory database systems: An
overview. IEEE Transactions on Knowledge and Data Engineering **4**(6), 509–
516 (1992)

[87] Gausemeier, J., Plass, C., Wenzelmann, C.: Zukunftsorientierte Un-
ternehmensgestaltung: Strategien, Geschäftsprozesse und IT-Systeme für die
Produktion von morgen. Hanser, München (2009)

[88] Geißler, R.: Die Sozialstruktur Deutschlands: Aktuelle Entwicklungen und theoretische Erklärungsmodelle ; Gutachten. Wiso-Diskurs. Abt. Wirtschafts- und Sozialpolitik der Friedrich-Ebert-Stiftung, Bonn (2010)

[89] German Engineering Federation (VDMA): Manufacturing Execution Systems (MES) Kennzahlen (2009)

[90] Ghosh, S.: Distributed Systems: An Algorithmic Approach, *Chapman & Hall/CRC computer and information science series*, vol. 13. Chapman & Hall/CRC, Boca Raton (2007). URL http://www.loc.gov/catdir/enhancements/fy0661/2006017600-d.html

[91] GmbH, F.: Award-winning leader in technology: World first: Complex event processing (CEP) with live cache (2011). URL http://www.forcam.co.uk/index.php?id=64&L=1

[92] Gong, Y., Janssen, M.: Measuring process flexibility and agility. In: T. Janowski, J. Davies (eds.) Proceedings of the 4th International Conference on Theory and Practice of Electronic Governance, pp. 173–182. ACM, New York and NY and USA (2010)

[93] Grauer, M., Karadgi, S., Metz, D.: Enhancement of transparency and adaptability by online tracking of enterprise processes. In: Wirtschaftinformatik Proceedings (WI 2011), pp. 282–291. Zurich and Switzerland (2011)

[94] Grauer, M., Karadgi, S., Metz, D., Schäfer, W.: An approach for real-time control of enterprise processes in manufacturing using a rule-based system. In: M. Schumann (ed.) Multikonferenz Wirtschaftsinformatik 2010, pp. 1511–1522. Universitätsverlag Göttingen c/o SUB Göttingen, Göttingen (2010)

[95] Grauer, M., Karadgi, S., Metz, D., Schäfer, W.: Real-Time Enterprise - Schnelles Handeln für produzierende Unternehmen. Wirtschaftsinformatik & Management (5), 13–15 (2010)

[96] Grauer, M., Karadgi, S., Metz, D., Schäfer, W.: Online monitoring and control of enterprise processes in manufacturing based on an event-driven architecture. In: M. Muehlen, J. Su (eds.) Lecture Notes in Business Information Processing (LNBIP), vol. 66, pp. 671–682. Springer, Berlin and Heidelberg (2011)

[97] Grauer, M., Karadgi, S., Müller, U., Metz, D., Schäfer, W.: Proactive control of manufacturing processes using historical data. In: R. Setchi, I. Jordanov, R.J. Howlett, L.C. Jain (eds.) Proceedings of 14th International Conference on Knowledge-Based and Intelligent Information and Engineering Systems (KES 2010), *Lecture Notes in Computer Science (LNCS)*, vol. 6277, pp. 399–408. Springer, Berlin (2010)

[98] Grauer, M., Metz, D., Karadgi, S., Schäfer, W.: Identification and assimilation of knowledge for real-time control of enterprise processes in manufacturing. In: C. Atzenbeck, O. Dini, M. Hitson, B. Jerman-Blazic (eds.) Proceedings of Second International Conference on Information, Process, and Knowledge Management (eKNOW 2010), pp. 13–16. IEEE (2010)

[99] Grauer, M., Metz, D., Karadgi, S., Schäfer, W., Reichwald, J.: Towards an IT-framework for digital enterprise integration. In: J. Kacprzyk, G.Q. Huang,

K.L. Mak, P.G. Maropoulos (eds.) Proceedings of the 6th CIRP-Sponsored International Conference on Digital Enterprise Technology (DET 2009), *Advances in Intelligent and Soft Computing (AISC)*, vol. 66, pp. 1467–1482. Springer, Berlin and Heidelberg (2010)

[100] Grauer, M., Müller, U., Metz, D., Karadgi, S., Schäfer, W.: About an architecture for integrated content-based enterprise search. In: Proceedings of 3rd International Conference on Information, Process, and Knowledge Management (eKNOW 2011), pp. 48–53. Gosier and Guadeloupe and France (2011)

[101] Grauer, M., Seeger, B., Metz, D., Karadgi, S., Schneider, M.: About adopting event processing in manufacturing. In: M. Cezon, Y. Wolfsthal (eds.) Service Wave 2010 Workshop Proceedings, *Lecture Notes in Computer Science (LNCS)*, vol. 6569, pp. 180–187. Springer, Berlin (2011)

[102] Groba, C., Braun, I., Springer, T., Wollschlaeger, M.: A service-oriented approach for increasing flexibility in manufacturing. In: Proceedings of International Workshop of Factory Communication Systems (WFCS), pp. 415–422. IEEE (2008)

[103] Gronau, N., Weber, E.: Management of knowledge intensive business processes. In: J. Desel, B. Pernici, M. Weske (eds.) Proceedings of Second International Conference on Business Process Management (BPM 2004), *Lecture Notes in Computer Science (LNCS)*, vol. 3080, pp. 163–178. Springer, Berlin (2004)

[104] Groover, M.P.: Fundamentals of Modern Manufacturing: Materials, Processes, and Systems, 3 edn. Wiley, Hoboken and NJ and USA (2006)

[105] Groover, M.P.: Automation, Production Systems, and Computer-Integrated Manufacturing, 3 edn. Prentice Hall, Upper Saddle River and NJ and USA (2008)

[106] Günther, H.O., Tempelmeier, H.: Produktion und Logistik, 8 edn. Springer, Berlin (2009)

[107] Gutenberg, E.: Die Produktion, 24 edn. Springer, Berlin (1983)

[108] Hammer, M., Champy, J.: Reengineering the Corporation: A Manifesto for Business Revolution. Harper Business, New York (1994)

[109] Han, J., Kamber, M.: Data Mining: Concepts and Techniques. Morgan Kaufmann Publishers, San Francisco (2001)

[110] Haunschild, L., Wolter, H.J.: Die volkswirtschaftliche Bedeutung der Familienunternehmen (2007). URL http://www.ifm-bonn.org/assets/documents/IfM-Materialien-172.pdf

[111] Heinrich, L.J.: Forschungsmethodik einer Integrationsdisziplin: Ein Beitrag zur Geschichte der Wirtschaftsinformatik. NTM International Journal of History & Ethics of Natural Sciences, Technology & Medicine **13**(2), 104–117 (2005)

[112] Helary, J., Raynal, M., Melideo, G., Baldoni, R.: Efficient causality-tracking timestamping. IEEE Transactions on Knowledge and Data Engineering **15**(5), 1239–1250 (2003)

[113] Henning, M.: The rise and fall of CORBA. Communications of the ACM **51**(8), 52–57 (2008)

[114] Hevner, A.R., March, S.T., Park, J., Ram, S.: Design science in information systems research. MIS Quarterly **28**(1), 75–105 (2004)

[115] Higgins, P., Le Roy, P., Tierney, L.: Manufacturing Planning and Control: Beyond MRP II. Chapman & Hall, London (1996)

[116] Hill, T.: Manufacturing Strategy: Text and Cases, 3 edn. McGraw-Hill, Boston (2000)

[117] Ho, L.T., Lin, G.C.I.: Critical success factor framework for the implementation of integrated-enterprise systems in the manufacturing environment. International Journal of Production Research **42**(17), 3731–3742 (2004)

[118] Hoffmann, D.: Fraunhofer Studie: MES ist noch kein Standard (2012). URL http://www.automotiveit.eu/fraunhofer-studie-mes-ist-noch-kein-standard/blickpunkt/id-0031804

[119] Holweg, M., Pil, F.K.: Evolving from value chain to value grid. MIT Sloan management review **47**(4), 72–80 (2006)

[120] Hugos, M.: Building the Real-time Enterprise: An Executive Briefing. John Wiley & Sons, Hoboken and NJ and USA (2005)

[121] IBM: J2EE vs. Microsoft .NET (s.a.). URL http://www-01.ibm.com/software/smb/na/J2EE_vs_NET_History_and_Comparison.pdf

[122] IMS International: Intelligent manufacturing systems: Global research and business innovation program. URL http://www.ims.org/

[123] Institut für Mittelstandsforschung (IfM): KMU-Definition des IfM Bonn (2002). URL http://www.ifm-bonn.org/index.php?id=89

[124] Institut für Mittelstandsforschung (IfM): Schlüsselzahlen der familienunternehmen nach ifm bonn (2006). URL http://www.ifm-bonn.org/index.php?id=905

[125] Institut für Mittelstandsforschung (IfM): Schlüsselzahlen des Mittelstands in Deutschland gemäß der KMU-Definition des IfM Bonn (2009). URL http://www.ifm-bonn.org/assets/documents/SZ-Unt&Ums&Besch_2004-2009&2010revSch_D_KMU_nach_IfM-Def.pdf

[126] International Electrotechnical Commission (IEC): Enterprise-control system integration - part 1: Models and terminology (2003-03)

[127] International Electrotechnical Commission (IEC): Enterprise-control system integration - part 3: Activity models of manufacturing operations management (2007-06)

[128] International Electrotechnical Commission (IEC): Function blocks – part 1: Architecture (IEC 65B/799/CDV:2011) (2011-12-05)

[129] International Organization for Standardization (ISO): Industrial automation systems - requirements for enterprise-reference architectures and methodologies (2000-06-01)

[130] Jacobsen, S.F., Eriksen, L., Kim, P.: Manufacturing 2.0: A Fresh Approach to Integrating Manufacturing Operations with DDVN (2010)

[131] Jammes, F., Smit, H.: Service-oriented architectures for devices - the SIRENA view. In: R. Schoop (ed.) Proceedings of 3rd IEEE International Conference on Industrial Informatics (INDIN), pp. 140–147. IEEE, Piscataway and NJ and USA (2005)

[132] Joachim, T., Vietor, M.: Kundenzentrierte Prozesse. In: A.W. Scheer, F. Abolhassan, W. Bosch (eds.) Real-Time Enterprise, pp. 27–42. Springer, Berlin (2003)

[133] Jörns, C.T.: Übergreifende koordination zum fulfillment im supply chain netzwerk. In: A.W. Scheer, F. Abolhassan, W. Bosch (eds.) Real-Time Enterprise, pp. 43–71. Springer, Berlin (2003)

[134] Kalmbach, P., Franke, R., Knottenbauer, K., Krämer, H., Schaefer, H.: Die Bedeutung einer wettbewerbsfähigen Industrie für die Entwicklung des Dienstleistungssektors: Eine Analyse der Bestimmungsgründe der Expansion industrienaher Dienstleistungen in modernen Industriestaaten (2003). URL http://www.bmwi.de/BMWi/Redaktion/PDF/C-D/ die-bedeutung-einer-wettbewerbsfaehigen-industrie-\ fuer-die-entwicklung-des-dienstleistungssektors, property=pdf,bereich=bmwi,sprache=de,rwb=true.pdf

[135] Kam, P.s., Fu, A.W.c.: Discovering temporal patterns for interval-based events. In: Y. Kambayashi, M. Mohania, A.M. Tjoa (eds.) Lecture Notes in Computer Science, vol. 1874, pp. 317–326. Springer, Berlin and Heidelberg (2000)

[136] Karadgi, S.: A framework for reinforcing existing monitoring and control of manufacturing processes based on enterprise performance measurement in real-time. Ph.D. thesis, University of Siegen, Siegen (s.a.)

[137] Karadgi, S., Metz, D., Grauer, M.: Real-time managerial accounting: A tool to monitor and control a manufacturing enterprise. In: International Workshop of Advanced Manufacturing and Automation (IWAMA 2012), p. (to appear) (2012)

[138] Karadgi, S., Metz, D., Grauer, M., Schäfer, W.: An event driven software framework for enabling enterprise integration and control of enterprise processes. In: A.E. Hassanien, A. Abraham, F. Marcelloni, H. Hagras, M. Antonelli, T.P. Hong (eds.) Proceedings of 10th International Conference on Intelligent Systems Design and Applications (ISDA), pp. 24–30. IEEE, Piscataway and NJ and USA (2010)

[139] Kärkkäinen, M., Ala-Risku, T., Främling, K.: Efficient tracking for short-term multi-company networks. International Journal of Physical Distribution & Logistics Management 34(7), 545–564 (2004)

[140] Karnouskos, S., Baecker, O., Souza, L.M.S.d., Spiess, P.: Integration of SOA-ready networked embedded devices in enterprise systems via a cross-layered web service infrastructure. In: Proceedings of Conference on Emerging Technologies and Factory Automation (ETFA), pp. 293–300. IEEE (2007)

[141] Karnouskos, S., Guinard, D., Savio, D., Spiess, P., Baecker, O., Trifa, V., Moreira Sa Souza, L.d.: Towards the real-time enterprise: Service-based

integration of heterogeneous SOA-ready industrial devices with enterprise applications. In: N. Bakhtadze, A. Dolgui (eds.) Proceedings of the 13th IFAC Symposium on Information Control Problems in Manufacturing (IN-COM 2009), pp. 2127–2132. IFAC (2009)

[142] Keller, G., Nüttgens, M., Scheer, A.W.: Semantische Prozeßmodellierung auf der Grundlage Ereignisgesteuerter Prozeßketten (EPK). In: A.W. Scheer (ed.) Veröffentlichungen des Instituts für Wirtschaftsinformatik. Saarbrücken (1992)

[143] Kemeny, Z., Ilie-Zudor, E., Szathmari, M., Monostori, L.: TraSer: An open-source solution platform for cross-company transparency in tracking and tracing. In: M.J. Chung, P. Misra, H. Shim (eds.) Proceedings of the 17th IFAC World Congress (IFAC 2008), pp. 4493–4498 (2008)

[144] Kendall, K.E., Kendall, J.E.: Systems Analysis and Design, 6 edn. Prentice Hall, Upper Saddle River and NJ (2004)

[145] Khabbazi, M.R., Ismail, M.Y., Ismail, N., Mousavi, S.A.: Modeling of traceability information system for material flow control data. Australian Journal of Basic and Applied Sciences 4(2), 208–216 (2010)

[146] Kjaer, A.P.: The integration of business and production processes. IEEE Control Systems Magazine 23(6), 50–58 (2003)

[147] Koestler, A.: The Ghost in the Machine. Macmillan, New York (1968)

[148] Kopetz, H.: Real-Time Systems: Design Principles for Distributed Embedded Applications, 2 edn. Springer, New York (2011)

[149] Korte, S., Rijkers-Defrasne, S., Zweck, A., Knackstedt, R., Lis, L.: Hybride Wertschöpfung: Ergebnisse des Förderschwerpunktes (2010). URL http://www.zukuenftigetechnologien.de/pdf/Band_89.pdf

[150] Krämer, J.: Continuous queries over data streams – semantics and implementation. Datenbanksysteme in Business, Technologie und Web (BTW) pp. 438–448 (2009)

[151] Krüger, J.J.: Productivity and structural change: A review of the literature. Journal of Economic Surveys 22(2), 330–363 (2008)

[152] Kurbel, K.: Produktionsplanung und -steuerung im Enterprise Resource Planning und Supply Chain Management, 6 edn. Oldenbourg, München (2005)

[153] Kusiak, A.: Data mining: Manufacturing and service applications. International Journal of Production Research 44(18-19), 4175–4191 (2006)

[154] Lamport, L.: Time, clocks, and the ordering of events in a distributed system. Communications of the ACM 21(7), 558–565 (1978)

[155] Langenscheidt, F., Bauer, U.: Lexikon der deutschen Weltmarktführer: Die Königsklasse deutscher Unternehmen in Wort und Bild, 1 edn. GABAL, Offenbach am Main (2010)

[156] Lee, J., Siau, K., Hong, S.: Enterprise integration with ERP and EAI. Communications of the ACM 46(2), 54–60 (2003)

[157] Leitao, P.: Agent-based distributed manufacturing control: A state-of-the-art survey. Engineering Applications of Artificial Intelligence 22(7), 979–991 (2009)

[158] Leitão, P., Vrba, P.: Recent developments and future trends of industrial agents. In: V. Mařík, P. Vrba, P. Leitão (eds.) Lecture Notes in Computer Science (LNCS), vol. 6867, pp. 15–28. Springer, Berlin and Heidelberg (2011)

[159] Louis, P.: Manufacturing Execution Systems: Grundlagen und Auswahl, 1 edn. Gabler, Wiesbaden (2009)

[160] Lovas, L.: Apama monitorscript by example (2008). URL http://apama.typepad.com/my_weblog/2008/02/apama-monitorsc.html

[161] Luckham, D.: The Power of Events: An Introduction to Complex Event Processing in Distributed Enterprise Systems. Addison-Wesley, Boston and Mass (2002)

[162] Luckham, D.: What's the difference between ESP and CEP? (2006). URL http://www.complexevents.com/2006/08/01/what%E2%80%99s-the-difference-between-esp-and-cep/

[163] Luckham, D., Schulte, R.: Event processing glossary - version 2.0 (2011). URL http://www.complexevents.com/wp-content/uploads/2011/08/EPTS_Event_Processing_Glossary_v2.pdf

[164] Luckham, D.C.: Event Processing for Business: Organizing the Real-Time Enterprise. Wiley, Hoboken and NJ and USA (2012)

[165] Lynch, R.L., Cross, K.F.: Measure up! Yardsticks for continuos improvement, 2 edn. Blackwell, Malden (1999)

[166] Mandal, A.K.: Introduction to Control Engineering: Modeling, Analysis and Design. New Age International (P) Ltd., Publishers, New Delhi (2006)

[167] Manenti, P.: Preliminary 2012 top 10 operations technology predictions: MESA european conference - cloudy with a chance for profits (November 9-10, 2011)

[168] Manufacturing Enterprise Solutions Association (MESA): Real-time enterprise strategic initiative guidebook: A MESA international guidebook (August 2008)

[169] Manufacturing Enterprise Solutions Association (MESA): About MESA (s.a.). URL http://mesa.org/en/aboutus/aboutmesa.asp

[170] March, S.T., Smith, G.F.: Design and natural science research on information technology. Decision Support Systems 15(4), 251–266 (1995)

[171] Mařík, V., Štěpánková, O., Krautwurmová, H., Luck, M. (eds.): Multi-Agent Systems and Applications II: 9th ECCAI-ACAI / EASSS 2001, AEMAS 2001, HoloMAS 2001 Selected Revised Papers, Lecture Notes in Computer Science (LNCS), vol. 2322. Springer, Berlin and Heidelberg (2002)

[172] Mattern, F.: Virtual Time and Global States of Distributed Systems, Bericht. Sonderforschungsbereich 124 VLSI Entwurfsmethoden und Parallelität, Fachbereich 10, Universität Saarbrücken. Universität Kaiserslautern, vol. 1988,38. Saarbrücken (1988)

[173] Maurer, C.: Measuring information systems agility: Construct definition and scale development. In: Proceedings of the Southern Association for Information Systems Conference, pp. 155–160. Atlanta and GA and USA (2010)

[174] McClellan, M., Weaver, D.: MESA model evolution: White paper #39 (2011)

[175] McGarry, K.: A survey of interestingness measures for knowledge discovery. The Knowledge Engineering Review **20**(1), 39–61 (2005)

[176] McGee, K.: Gartner updates its definition of real-time enterprise (2004)

[177] Meldal, S., Sankar, S., Vera, J.: Exploiting locality in maintaining potential causality. In: Proceedings of the 10th Annual ACM Symposium on Principles of Distributed Computing, pp. 231–239. ACM Press, New York and NY (1991)

[178] Mendes, M.R., Bizarro, P., Marques, P.: A performance study of event processing systems. In: R. Nambiar (ed.) Performance Evaluation and Benchmarking, pp. 221–236. Springer, Berlin and Heidelberg and New York and NY (2009)

[179] Merriam Webster Learner's Dictionary: causality (2012). URL http://www.learnersdictionary.com/search/causality

[180] Merriam Webster Learner's Dictionary: event (2012). URL http://www.learnersdictionary.com/search/event

[181] Metz, D.: Integriertes Produktionsmanagement in einer Aluminiumgießerei (24.03.2012)

[182] Metz, D., Karadgi, S., Grauer, M.: A process model for establishment of knowledge-based online control of enterprise processes in manufacturing. International Journal on Advances in Life Sciences **2**(3 & 4), 188–199 (2010)

[183] Meyer, C.: Expert voice: Christopher Meyer on the accelerating enterprise (2002). URL http://www.cioinsight.com/index2.php?option=content&do_pdf=1&id=881505

[184] Michelson, B.: Event-driven architecture overview: Event-driven SOA is just part of the EDA story (2006). URL http://www.omg.org/soa/Uploaded%20Docs/EDA/bda2-2-06cc.pdf

[185] Microsoft Developer Network (MSDN): Constrained execution regions (2012). URL http://msdn.microsoft.com/en-us/library/ms228973.aspx

[186] Microsoft Developer Network (MSDN): Monitor class (2012). URL http://msdn.microsoft.com/en-us/library/system.threading.monitor%28v=vs.100%29.aspx

[187] Molina, A., Panetto, H., Chen, D., Whitman, L., Chapurlat, V., Vernadat, F.: Enterprise integration and networking: challenges and trends. Studies in Informatics and Control **16**(4), 353–368 (2007)

[188] Mollenkopf, A., Tirelli, E.: Applying Drools Fusion complex event processing (CEP) for real-time intelligence (2009). URL http://www.redhat.com/f/pdf/jbw/amollenkopf_430_applying_drools.pdf

[189] Mönch, L.: Autonome und kooperative Steuerung komplexer Produktionsprozesse mit Multi-Agenten-Systemen. WIRTSCHAFTSINFORMATIK **48**(2), 107–119 (2006)

[190] Mühl, G., Fiege, L., Pietzuch, P.: Distributed Event-Based Systems. Springer, Berlin (2006)

[191] Nance, R.E.: A history of discrete event simulation programming languages (1993)

[192] National Science Foundation: Cyber-physical systems (CPS): Program solicitation (2009). URL http://www.nsf.gov/pubs/2008/nsf08611/nsf08611.pdf

[193] Niehaves, B.: Epistemological perspectives on multi-method information systems research. In: 13th European Conference on Information Systems (ECIS 2005) (2005)

[194] Object Management Group: Event service (EVNT): Version 1.2 (2004). URL http://www.omg.org/spec/EVNT/

[195] Offutt, J., Abdurazik, A.: Generating tests from uml specifications. In: R. France, B. Rumpe (eds.) Lecture Notes in Computer Science, vol. 1723, pp. 416–429. Springer, Berlin and Heidelberg (1999)

[196] Ohm & Häner Metallwerk GmbH & Co. KG: Wir über uns. URL http://www.ohmundhaener.de/index.php?option=com_content&view=article&id=62&Itemid=71&lang=de

[197] van Oosterhout, M., Waarts, E., van Heck, E., van Hillegersberg, J.: Business agility: Need, readiness and alignment with IT strategies. In: K. Desouza (ed.) Agile Information Systems, pp. 52–69. Butterworth-Heinemann, Amsterdam (2007)

[198] OPC Foundation: Cooperation of OPC foundation and PLCopen interoperability of the new generation (s.a.). URL http://www.opcfoundation.org/DownloadFile.aspx/Brochures/OPC-UA-CollaborationPLCopen.pdf?RI=804

[199] Österle, H. (ed.): Gestaltungsorientierte Wirtschaftsinformatik: Ein Plädoyer für Rigor und Relevanz. Infowerk, Nürnberg and Germany (2010). URL http://www.alexandria.unisg.ch/export/DL/213293.pdf

[200] Ovacik, I.M., Uzsoy, R.: Decomposition Methods for Complex Factory Scheduling Problems. Kluwer Academic Publishers, Boston (1997)

[201] Panetto, H., Molina, A.: Enterprise integration and interoperability in manufacturing systems: Trends and issues. Computers in Industry 59(7), 641–646 (2008)

[202] Pessoa, R.M., Silva, E., van Sinderen, M., Quartel, D.A., Pires, L.F.: Enterprise interoperability with SOA: A survey of service composition approaches. In: 12th Enterprise Distributed Object Computing Conference Workshops (EDOCW), pp. 238–251. IEEE, Piscataway and NJ and USA (2008)

[203] Pham, D.T., Afify, A.A.: Machine-learning techniques and their applications in manufacturing. Journal of Engineering Manufacture 219(5), 395–412 (2005)

[204] Pinedo, M.: Scheduling: Theory, Algorithms, and Systems, 2 edn. Prentice Hall, Upper Saddle and NJ (2002)

[205] Piontek, J.: Controlling, 3 edn. Oldenbourg, München (2005)

[206] Plattner, H., Zeier, A.: In-Memory Data Management: An Inflection Point for Enterprise Applications. Springer, Berlin Heidelberg (2011)

[207] Porter, M.E.: Competitive advantage: Creating and Sustaining Superior Performance. Free Press, New York (1998). URL http://www.loc.gov/catdir/bios/simon051/98009581.html

[208] Ragunathan, R., Lee, I., Sha, L., Stankovic, J.: Cyber-physical systems: The next computing revolution. In: S.S. Sapatnekar (ed.) Proceedings of 47th ACM/IEEE Design Automation Conference (DAC 2010), pp. 731–736. IEEE, Piscataway and NJ (2010)

[209] Rammer, C., Aschhoff, B., Crass, D., Doherr, T., Hud, M., Köhler, C., Peters, B., Schubert, T., Schwiebacher, F.: Innovationsbericht der deutschen Wirtschaft: Indikatorenbericht zur Innovationserhebung 2011 (2012). URL http://ftp.zew.de/pub/zew-docs/mip/11/mip_2011.pdf

[210] Rao, T.V.: Metal Casting: Principles and Practice. New Age International Publishers, New Delhi (2003)

[211] Rapide Design Team: Guide to the rapide 1.0 language reference manual (1997). URL http://complexevents.com/stanford/rapide/lrms/overview.ps

[212] Raskino, M., McGee, K.: Addressing the CEO's demand for real-time warnings (2004)

[213] Reichwald, J.: Modell-getriebene Unterstützung der Workflow-Abbildung in Service-orientierten Software-Umgebungen. Ph.D. thesis, Universität Siegen, Siegen (2009)

[214] Riege, C., Saat, J., Bucher, T.: Systematisierung von Evaluationsmethoden in der gestaltungsorientierten Wirtschaftsinformatik. In: J. Becker, H. Krcmar, B. Niehaves (eds.) Wissenschaftstheorie und gestaltungsorientierte Wirtschaftsinformatik, pp. 69–86. Physica-Verlag, Heidelberg and Germany (2009)

[215] Robra-Bissantz, S.: Wissenschaft und Forschung in der Wirtschaftsinformatik (11.03.2008)

[216] Rommelspacher, J.: Ereignisgetriebene Architekturen. WIRTSCHAFTSINFORMATIK 50(4), 314–317 (2008)

[217] Rommelspacher, J.: Modelling complex events with event-driven process chains. In: W. Hesse, A. Oberweis (eds.) Proceedings of the Third AIS SIGSAND European Symposium on Analysis, Design, Use and Societal Impact of Information Systems, pp. 79–82. Bonn (2008)

[218] Rosenberg, M.: Sectors of the economy: Primary, secondary, tertiary, quaternary, and quinary (2007). URL http://geography.about.com/od/urbaneconomicgeography/a/sectorseconomy.htm

[219] Roth, R., Tilkov, S.: Ereignis-getriebene Architekturen: ein Überblick. OBJEKTspektrum (2), 26–27 (2006)

[220] van Roy, P.: Programming paradigms for dummies: What every programmer should know. In: G. Assayag, A. Gerzso (eds.) New computational paradigms for computer music, pp. 9–47. Delatour and IRCAM, Sampzon and Paris (2009)

[221] Ryu, K., Jung, M.: Agent-based fractal architecture and modelling for developing distributed manufacturing systems. International Journal of Production Research **41**(17), 4233–4255 (2003)

[222] Sauer, O., Sutschet, S.: Agent-based control. Computing and Control Engineering **17**(3), 32 (2006)

[223] Scheer, A.W.: Business Process Engineering: Reference Models for Industrial Enterprises, 2 edn. Springer, Berlin (1994)

[224] Scheer, A.W.: 20 Jahre Gestaltung industrieller Geschäftsprozesse. Industrie Management **20**(1), 11–18 (2004)

[225] Scheer, A.W., Abolhassan, F., Bosch, W. (eds.): Real-Time Enterprise: Mit beschleunigten Managementprozessen Zeit und Kosten sparen. Springer, Berlin (2003)

[226] Scholz-Reiter, B., Freitag, M.: Autonomous processes in assembly systems. CIRP Annals - Manufacturing Technology **56**(2), 712–729 (2007)

[227] Schonenberg, H., Mans, R., Russell, N., Mulyar, N., van der Aalst, W.M.P.: Process flexibility: A survey of contemporary approaches. In: J.L.G. Dietz, A. Albani, J. Barjis (eds.) Advances in Enterprise Engineering, pp. 16–30. Springer, Berlin (2008)

[228] Schönherr, M.: Enterprise architecture frameworks. In: S. Aier, M. Schönherr (eds.) Enterprise Application Integration – Serviceorientierung und nachhaltige Architekturen, *Enterprise Architecture*, vol. 2, pp. 3–48. Gito, Berlin (2004)

[229] Schulte, R.: The growing role of events in enterprise applications (July 2003)

[230] Schulte, R.: A real-time enterprise is event-driven (September 2002)

[231] Seiriö, M.: So why not use an in-memory db instead? (2009). URL http://rulecore.com/CEPblog/?p=257

[232] Shahbaz, M., Srinivas, M., Harding, J.A., Turner, M.: Product design and manufacturing process improvement using association rules. Journal of Engineering Manufacture **220**(2), 243–254 (2006)

[233] Shannon, C.: A mathematical theory of communication. The Bell System Technical Journal **27**, 379–423, 623–656 (1948). URL http://cm.bell-labs.com/cm/ms/what/shannonday/shannon1948.pdf

[234] Simon, H.A.: The Sciences of the Artificial, 3 edn. MIT Press, Cambridge and Mass (1996)

[235] SIRI: Service interface for real time information CEN/TS 15531 (prCEN/TS-OO278181) (2011). URL http://www.kizoom.com/standards/siri/

[236] Socrades: Welcome to SOCRADES 2006-2009 (s.a.). URL http://www.socrades.eu/

[237] Souza, L.M.S., Spiess, P., Guinard, D., Köhler, M., Karnouskos, S., Savio, D.: SOCRADES: A web service based shop floor integration infrastructure. In: C. Floerkemeier, M. Langheinrich, E. Fleisch, F. Mattern, S.E. Sarma (eds.) Proceedings of the First International Conference on The Internet of

Things, *Lecture Notes in Computer Science (LNCS)*, vol. 4952, pp. 50–67. Springer, Berlin and Heidelberg (2008)

[238] Stadtler, H., Kilger, C. (eds.): Supply Chain Management and Advanced Planning: Concepts, Models, Software, and Case Studies, 4 edn. Springer, Berlin (2008)

[239] Stahl, B.C.: The ideology of design: A critical appreciation of the design science discourse in information systems and Wirtschaftsinformatik. In: J. Becker, H. Krcmar, B. Niehaves (eds.) Wissenschaftstheorie und gestaltungsorientierte Wirtschaftsinformatik, pp. 111–132. Physica-Verlag, Heidelberg and Germany (2009)

[240] Stark, J.: Product Lifecycle Management: 21st Century Paradigm for Product Realisation. Springer, London (2011)

[241] Sundermeyer, K., Bussmann, S.: Einführung der Agententechnologie in einem produzierenden Unternehmen - Ein Erfahrungsbericht. WIRTSCHAFTSINFORMATIK **43**(2), 135–142 (2001)

[242] Tanenbaum, A.S., Van Steen, M.: Distributed Systems: Principles and Paradigms, 2 edn. Pearson/Prentice Hall, Upper Saddle River and NJ (2007)

[243] Taylor, H., Yochem, A., Phillips, L., Martinez, F.: Event-Driven Architecture: How SOA Enables the Real-Time Enterprise. Addison-Wesley, Upper Saddle River and NJ and USA (2009)

[244] Tewari, A.: Modern Control Design with MATLAB and SIMULINK. John Wiley & Sons, Chichester (2005)

[245] Thammatutto, C.: Improving production efficiency through lean flexible cellular manufacturing simulation. In: Proceedings of IEEE International Conference on Quality and Reliability, pp. 322–326. IEEE (2011)

[246] Tharumarajah, A., Wells, A., Nemes, L.: Comparison of emerging manufacturing concepts. In: IEEE International Conference on Systems, Man, and Cybernetics, pp. 325–331. IEEE (1998)

[247] The International Society of Automation (ISA): Enterprise control system integration part 3: Activity models of manufacturing operations management (2005)

[248] Thiel, K.: MES - Integriertes Produktionsmanagement: Leitfaden, Marktübersicht und Anwendungsbeispiele. Hanser, Carl, München (2011)

[249] TIBCO: Introducing TIBCO BusinessEvents 3.0: The Leader in Complex Event Processing (2012). URL http://www.tibco.com/products/business-optimization/complex-event-processing/businessevents/default.jsp

[250] Toyryla, I.: Realising the Potential of Traceability: A Case Study Research on Usage and Impacts of Product Traceability. Finnish Academy of Technology, Espoo and Finland (1999)

[251] Velamuri, V.K.: Hybrid value creation (s.a.). URL http://www.hybridvaluecreation.com/definitions.html

[252] Venohr, B.: The power of uncommon common sense management principles. the secret recipe of German Mittelstand companies - lessons for large and small companies: Presentation, 2nd global Drucker Forum, Vienna, Austria,

November 18-19, 2010 (2010). URL http://www.berndvenohr.de/ download/vortraege/101117_B_Drucker%20Forum_FIN.pdf

[253] Venohr, B., Meyer, K.: The German miracle keeps running: How Germany's hidden champions stay ahead in the global economy (2007)

[254] Venohr, B., Meyer, K.: Uncommon common sense. Business Strategy Review **20**(1), 39–43 (2009)

[255] Verein Deutscher Ingenieure / Kompetenzfeld Informationstechnik: Manufacturing Execution Systems (MES), *VDI-Richtlinien*, vol. 5600, Blatt 1. VDI-Verlag, Düsseldorf (2006)

[256] Verein Deutscher Ingenieure / Kompetenzfeld Informationstechnik: Manufacturing Execution Systems (MES): Logic interfaces for machine and plant control, *VDI-Richtlinien*, vol. VDI 5600, Blatt 3 Entwurf. VDI-Verlag, Düsseldorf (2011)

[257] Verein Deutscher Ingenieure / Kompetenzfeld Informationstechnik: Manufacturing Execution Systems (MES): Support of production systems by MES, *VDI-Richtlinien*, vol. VDI 5600, Blatt 4 Entwurf. VDI-Verlag, Düsseldorf (2011)

[258] Verein Deutscher Ingenieure / Kompetenzfeld Informationstechnik: Manufacturing Execution Systems (MES): Cost effectiveness, *VDI-Richtlinien*, vol. VDI 5600, Blatt 2 Entwurf. VDI-Verlag, Düsseldorf (2012)

[259] Verilog: Verilog (s.a.). URL http://www.mikrocontroller.net/ articles/Verilog

[260] Vidackovic, K., Renner, T., Rex, S.: Marktübersicht Real-Time-Monitoring-Software: Event-Processing-Tools im Überblick. Fraunhofer-Verlag, Stuttgart (2010)

[261] Vijayaraghavan, A.: MTConnect for realtime monitoring and analysis of manufacturing enterprises: Presented at international conference on digital enterprise technology (DET 2009) (December 14-16, 2009)

[262] Wadhwa, S., Rao, K.S.: Flexibility and agility for enterprise synchronization: Knowledge and innovation management towards flexagility. Studies in Informatics and Control **12**(2), 111–128 (2003)

[263] Wagner, T., Göhner, P., Urbano, P.G.A.d.: Softwareagenten - Einführung und Überblick über eine alternative Art der Softwareentwicklung: Teil 1: Agentenorientierte Softwareentwicklung. atp **45**(10), 48–57 (2003)

[264] Wallau, F., Adenhäuser, C., Kayser, G.: BDI-Mittelstandspanel: Ergebnisse der Online-Mittelstandsbefragung (2006). URL http://www. ifm-bonn.org/assets/documents/IfM-Materialien-169. pdf

[265] Walzer, K., Rode, J., Wunsch, D., Groch, M.: Event-driven manufacturing: Unified management of primitive and complex events for manufacturing monitoring and control. In: Proceedings of International Workshop of Factory Communication Systems (WFCS), pp. 383–391. IEEE (2008)

[266] Weske, M., van der Aalst, W., Verbeek, H.: Advances in business process management. Data & Knowledge Engineering **50**(1), 1–8 (2004)

[267] Wiendahl, H.P., ElMaraghy, H.A., Nyhuis, P., Zäh, M.F., Wiendahl, H.H., Duffie, N., Brieke, M.: Changeable manufacturing - classification, design and operation. CIRP Annals - Manufacturing Technology **56**(2), 783–809 (2007)

[268] Wiendahl, H.P., Heger, C.L.: Justifying changeability: A methodical approach to achieving cost effectiveness. International Journal for Manufacturing Science and Production **6**(1-2), 33–40 (2005)

[269] Wiener, N.: Cybernetics or Control and Communication in the Animal and the Machine. Massachusetts Institute of Technology Press, New York (1948)

[270] Williams, T.J., Rathwell, G.A., Li, H.: A handbook on master planning and implementation for enterprise integration: Based on the purdue enterprise reference architecture and the Purdue methodology (1996)

[271] Wochinger, T., Weskamp, M.: Effizient-Werkzeug für Industriebetriebe: Einsatz und Nutzen von MES. IT & Production (Wissen Kompakt, Manufacturing Execution Systems (MES), 2012/13), 10–13 (2012). URL http://www.it-production.com/media_container/wissen_kompakt/ITP_MES_2012.pdf

[272] Wollert, J.F.: Einführung. In: H.J. Gevatter (ed.) Handbuch der Mess- und Automatisierungstechnik in der Produktion, pp. 475–484. Springer, Berlin (2006)

[273] Wulf, W., Cohen, E., Corwin, W., Jones, A., Levin, R., Pierson, C., Pollack, F.: HYDRA: The kernel of a multiprocessor operating system. Communications of the ACM **17**(6), 337–345 (1974)

[274] Zäh, M. (ed.): Proceedings of 3rd International Conference on Changeable, Agile, Reconfigurable and Virtual Production (CARV 2009): Munich, Germany, October 5-7, 2009. Utz, München (2009)

[275] Zhang, Y.H., Dai, Q.Y., Zhong, R.Y.: An extensible event-driven manufacturing management with complex event processing. International Journal of Control and Automation **2**(3), 1–12 (2009)

[276] Zoitl, A., Strasser, T., Hall, K., Staron, R., Sünder, C., Favre-Bulle, B.: The past, present, and future of IEC 61499. In: V. Mařík, V. Vyatkin, A.W. Colombo (eds.) Holonic and Multi-Agent Systems for Manufacturing, *Lecture Notes in Computer Science (LNCS)*, vol. 4659, pp. 1–14. Springer, Berlin and Heidelberg (2007)